Warning

The author and publisher cannot be held responsible in the event of an identification error made by the reader while gathering mushrooms. There are basic guidelines that can be applied when foraging for wild mushrooms throughout the world in order to prevent accidentally picking poisonous and deadly fungi. For example, follow these recommendations from the French Ministry of Health:

- Do not pick a mushroom if you have even the slightest doubt about its condition or identification. Certain highly toxic mushrooms strongly resemble edible species.

- Only gather specimens that are in good condition, and remove the entire mushroom (stem, base, and cap) for identification.

- Avoid polluted areas (roadsides, industrial zones, landfills) because mushrooms tend to accumulate pollutants; if you are the least bit unsure, throw the mushroom away.

- Collect the mushrooms one by one in a crate or box, but never in a plastic bag, which accelerates decay.

- Wash your hands well when you have finished harvesting.

- Separate the harvested mushrooms by species.

- Do not consume what you have harvested before having it checked by a specialist. You can consult a mycology organization in your area. In North America, the North American Mycological Association maintains a list of local mycological organizations. In the United Kingdom, contact the British Mycological Society for information. Set aside a portion of your harvest, uncooked and refrigerated, to show the specialist, most especially the first time you're eating a species for identification in case of poisoning symptoms.

- Keep mushrooms in the refrigerator and eat them no more than two days after your harvest.

- Consume mushrooms in reasonable amounts after cooking them sufficiently. In all but exceptional cases, never eat them raw.

In the event that you experience one or several of the symptoms associated with the consumption of wild mushrooms (shivering, vertigo, vision problems, nausea, vomiting, etc.), call a poison control center immediately. Symptoms may appear up to twelve hours after consumption. Record the time of your most recent meal or meals and make a note of when the first symptoms appeared. Then set aside the rest of your harvest for identification.

10 9 8 7 6 5 4

Cover design by Mona Lin

Interior illustrations by Julien Norwood

Library of Congress Cataloging-in-Publication Data is available on file.

Print ISBN: 978-1-5107-4067-9

eBook ISBN: 978-1-5107-4068-6

Printed in China

The Ultimate Guide to Mushrooms

The Ultimate Guide to Mushrooms

How to Identify and Gather Over 200 Species Throughout North America and Europe

BY GUILLAUME EYSSARTIER
ILLUSTRATIONS BY JULIEN NORWOOD

TRANSLATED BY GRACE McQUILLAN

Skyhorse Publishing

Correctly Identifying Each Mushroom

Illustrated panel allowing for easy identification of the mushroom in its different stages

Cross-referencing with other mushrooms for comparison

Additional information and anecdotes

Common name

Latin name

King
bolete

Boletus edulis Bull. : Fr.

The king bolete, reported from across North America and the UK, is often confused with the summer bolete (p. 127), which luckily is also edible. The king bolete's smooth and sometimes greasy cap, which loses color progressively approaching the edge, is nevertheless its classic feature. It is highly variable in color, and young specimens are sometimes completely white. There has been a significant amount of squabbling over the supposed differences in flavor linked to a king bolete's habitat: a number of people would say growing under conifers is best, while others insist that it does not deserve to go in the pan unless it grows under deciduous trees. In this area it is difficult to be objective. Mycologists continue their struggle to determine the true species in this complex. For now, in eastern North America, it is called *B. chippewanensis*.

In the kitchen

This bolete can be prepared in many ways: in a creamy sauce, in a salad (still always cooked), grilled, in a soufflé, and even in a dessert, caramelized with vanilla custard. To store it, choose specimens that are young and firm with tubes that are still whitish or yellow. It stores well dried, frozen, or pickled. It is better not to overwash it and it should never be soaked, because this might take away some of the flavor. Removing impurities with a knife and wiping it down with a clean towel is usually enough to get it ready for cooking.

Warning

The bitter bolete (p. 221) strongly resembles the king bolete. It is not poisonous, but its unbearable bitterness may leave you with no other choice but to throw a whole dish into the trash. Knowing how to identify it, therefore, is very important: its tubes rapidly turn pink, it has a rather brownish-pink spore print, its stem has a visible and pronounced dark web, and it grows under conifers. If you are not certain, taste a small piece.

Bring me my ceps!

The king bolete's reputation speaks for itself, and this mushroom arouses a great deal of passion in some regions. Harvesters and landowners have exchanged countless numbers of slashed tires, paint scratches, and even punches and gunshots over a few pounds of this mushroom. To remedy this significant problem, some municipalities have passed decrees to regulate cep harvesting, but those only apply to public lands: in private woods, the war rages on. In some areas, landowners organize themselves into societies, handing out harvesting maps and hiring guards to ensure that their property is respected and not completely ravaged by hordes of gatherers.

- ◉ **Cap:** *up to 10 inches (25 cm) wide (sometimes more), typically glossy and a little greasy to the touch, varying from whitish to dark brown, lighter and lighter in color approaching the edge.*
- ◉ **Tubes:** *thin and crowded, cream then yellow and finally olive-yellow; spore print olive-brown.*
- ◉ **Stem:** *fleshy and often ventricose, rather pale then ochre, a more or less pronounced web that is often limited to the top of the stem.*
- ◉ **Flesh:** *firm, white, a little brownish-pink just beneath the cap's skin; sweet flavor and very pleasant odor.*
- ◉ **Habitat:** *under deciduous trees and coniferous trees.*
- ◉ **When to harvest:** *Spring or autumn (rare in summer).*

cap-stem-tubes | **129**

Illustrated panel allowing for easy identification of the mushroom in its different stages

Identification index

Illustration of the mushroom in its habitat

Edible

General shape of the mushroom

Cooking instructions and flavor profiles

Potential sources of confusion

🍴 Edible

☠ Deadly

🍄 Inedible or Poisonous

🐙 Bizarre

Contents

▲ *Amethyst deceiver* (*Laccaria amethystina*)

Introduction

What Is a Mushroom?

▲ Surface of gills producing spores

The answer to this question appears so obvious that it may seem a ridiculous one to ask: a mushroom is "a plant without leaves or flowers, generally found in the form of a stem topped with a cap, consisting of numerous species, edible or poisonous, that grow rapidly, particularly in humid places." This definition, which to a mycologist sounds like it comes from an eighteenth-century encyclopedia, was actually taken from the 2011 edition of a very famous dictionary, a fact that is surprising considering that biologists have known since the 1960s that mushrooms have very little to do with plants and are, contrary to most expectations, more closely related to animals. Their cells, for instance, are bordered by walls made of chitin—which is also found in the "shells" of arthropods like insects and crustaceans—and not the pectin and cellulose seen in plant cells. Like animal cells, these cells function using energy stored in the form of glycogen, not starch. In short, mushrooms are not plants, and today are grouped in a separate kingdom: Kingdom Fungi.

▲ Fly agaric (Amanita muscaria). *Carpophore: this is the visible portion of basidiomycete mushrooms, consisting of a cap and a stem (see p. 16)*

▲ *King bolete* (Boletus edulis)

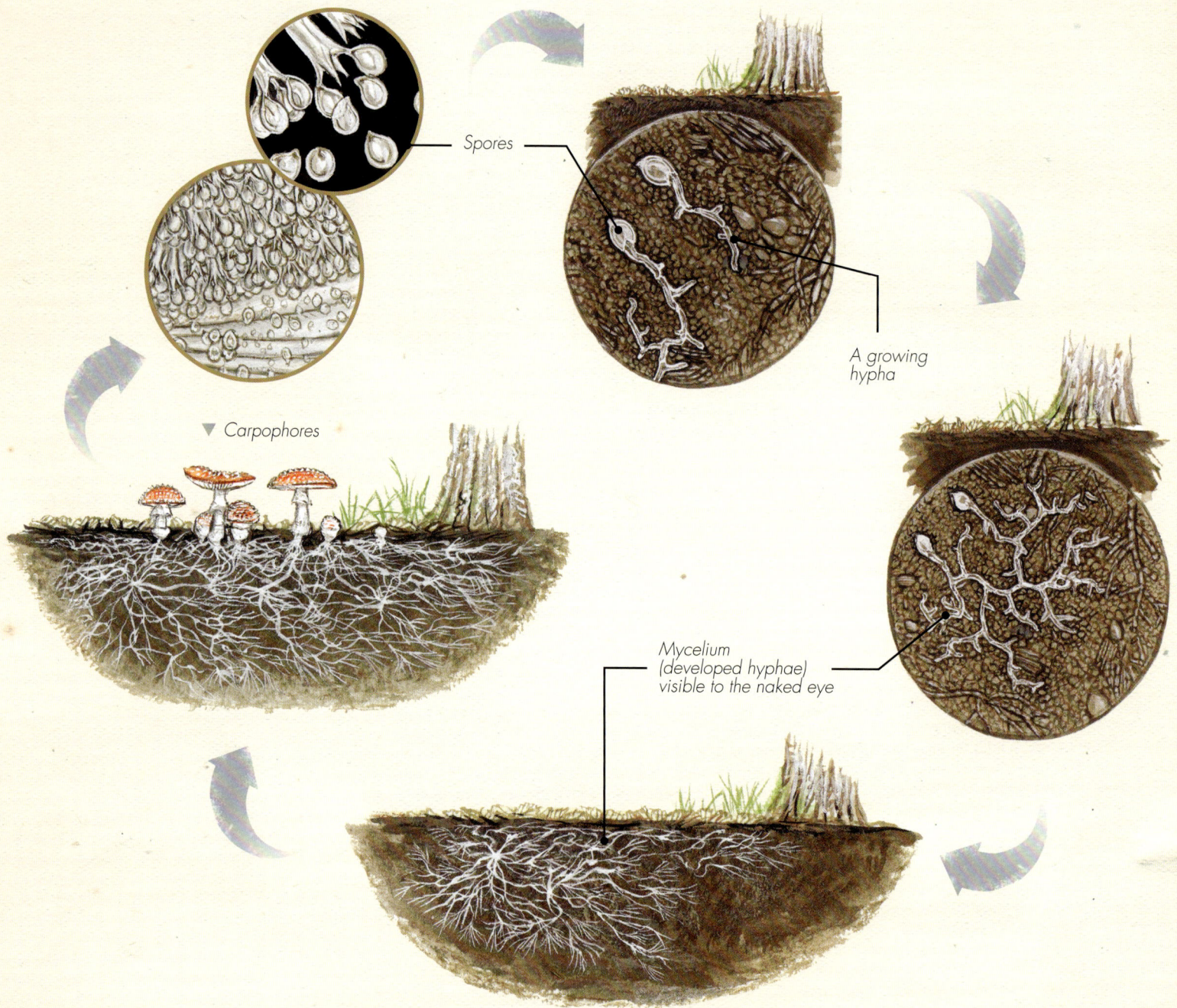

Spores

A growing hypha

Carpophores

Mycelium (developed hyphae) visible to the naked eye

Life Cycle: Fly agaric

Many chosen . . .

How many mushrooms, as we have defined them here, are in existence on Earth's surface? No one at present is able to answer that question. In North America alone, it is estimated that more than 20,000 species exist. An entire world . . . and the inventory is still far from being completed. Biologists today, for instance, are discovering a variety of species that until now no one had imagined existed: microscopic mushrooms that live inside plant tissues and are called "endophytes" (from the Greek *endon*, "within," and *phytos*, "plant"). Mycologists decided to study the endophyte mushrooms in a palm tree and counted no fewer than 418 species, 140 of which were inside a single leaf! Seventy-five percent of these species were "new to science," meaning they had never before been studied or described. No need to go probing the innermost layers of our good old Scotch pine (*Pinus sylvestris*) to find mushrooms, though: this tree is home to 892 different mushrooms, from those that grow alongside its roots (see *Mycorrhizal mushrooms*) to those that thrive on its wood and decomposing needles. Among these species, 186 are exclusively dependent on the Scotch pine and will not merge with any other tree.

The diversity of mushrooms on our planet is widely debated: it varies between 1.5 million (a very low approximation that we already know to be inaccurate) and over 30 million, and it is estimated that we are probably only aware of about three percent of the mushrooms present on the surface of the Earth.

. . . And a few left out

The current definition of Kingdom Fungi excludes certain groups of organisms that have traditionally been thought of as mushrooms. The potato blight (*Phytophthora infestans*) and brown rot (*Plasmopara viticola*), for instance, are now classified near certain seaweeds. But let's pause for a moment to talk about a strange group that you are likely to encounter when out for a walk (as long as you are paying attention, for its members are often small): the myxomycetes, the slime molds. These timid but charming organisms, today classified in their own group, the mycetozoa, grow on pretty much anything in any season, as long as it is not too cold. Instead of producing a filament (the hypha) like true mushrooms when they germinate, their spores—the microscopic cells of sexual reproduction—release a sort of amoeba that moves around on the substrate by crawling or propelling itself with flagella and feeds itself by absorbing bacteria and decomposing organic particles (the phenomenon of phagocytosis). When the site becomes unfavorable (because it dries up, for example), these amoebae gather together and fuse to create a "super amoeba" called a plasmodium that will eventually settle down and differentiate into smaller individual structures of various shapes and colors depending on the species. These structures will then produce spores, completing the life cycle. The plasmodium is also mobile and can move up to dozens of inches at a time before it finds a resting place.

▼ *Armillaria*

▲ *Orange birch bolete* (Leccinum versipelle)

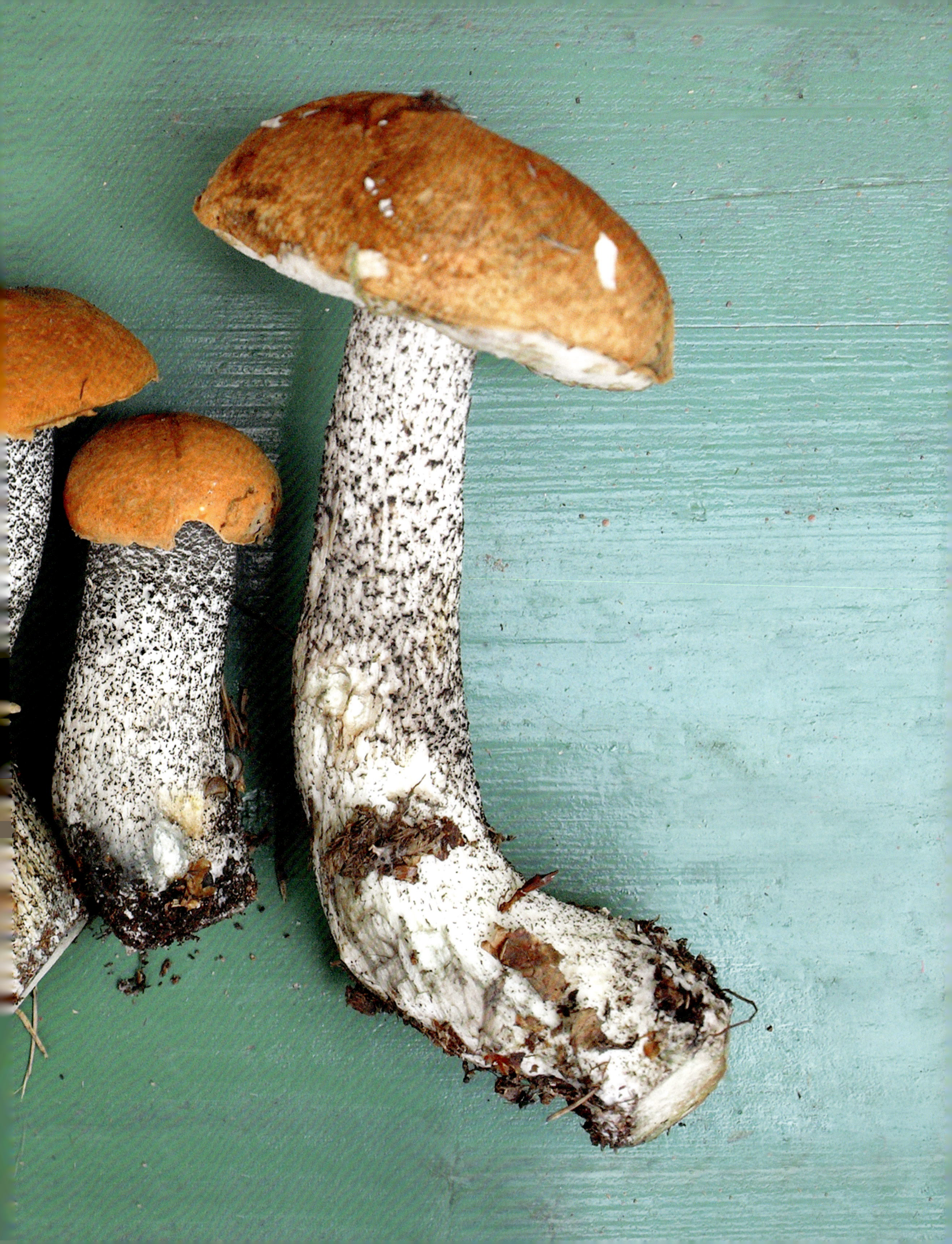

Naming
Mushrooms

All classified organisms on Earth today possess a Latin name, and mushrooms are no exception. These Latin names, made up of two terms, may seem cumbersome, but they are the only guarantee of universal communication: all mycologists in the world use them, whereas very few specialists would know more than a handful of the dozens of common names for the chanterelle around the world. Let's examine the Latin name for the chanterelle, *Cantharellus cibarius*. The Latin name for all of the mushrooms in this group begins with *Cantharellus*: this is what is called the genus name. It's a sort of "family name." However, only the "true" chanterelle bears the full name *Cantharellus cibarius*, which is the name of the species. The term *cibarius*, meaning "edible," is called the "specific epithet" and is used to distinguish this well-known mushroom from others.

Common names are sometimes misleading: the confusion they cause can be innocent when they suggest proximities between species that are not at all related (in the case of *Hydnum*, for example, which belong in fact to several different families), but it can also prove to be very dangerous. For example, for a long time, the French have called the deadly "Cortinaire couleur de rocou" (fool's webcap, p. 233) the "Cortinaire des montagnes," substantiating the idea that it grows only in mountainous areas and potentially endangering amateurs at lower elevations who are not expecting to encounter it. Also in French, the "gyromitre" (false morel, p. 248) is sometimes still known as the gyromitre "comestible," meaning "edible," even though its extreme toxicity leaves very little room for doubt; warning mushroom enthusiasts about this is imperative for obvious reasons.

▲ *Fly agaric* (Amanita muscaria)

▲ *Chanterelle* (Cantharellus cibarius)

The Different
Mushroom Groups

Mushroom classification has evolved significantly in the last twenty years. Modern methods in genetics and molecular biology have allowed us to move away from comparing organisms based solely on appearance and instead begin making direct comparisons using their genetic information—their DNA, or deoxyribonucleic acid—replacing the static view we used to have of relationships between species with a dynamic view that reflects the evolutions of those species. Though examining the details of this new kind of classification—one that is still evolving, in fact—is not the purpose of this book, we still need to provide a few major reference points as we begin to focus on the mushrooms we are interested in here: those that are edible and those that are toxic, all of which fall into two large groups.

The basidiomycetes

Included in this group are all mushrooms whose **spores** are formed on the *exterior* of cells called **basidia**. These cells are situated side by side to form an even fertile layer called the **hymenium** (see *Mushroom Anatomy* after this). These spores are usually violently expelled as soon as they are ripe (see *The Importance of Spore Dispersal*). Among the basidiomycetes are a large number of edible and toxic mushrooms, including all mushrooms that have gills, tubes, or teeth under their caps. The others, like morels and truffles, for example, are *ascomycetes*.

The ascomycetes

In this case, the spores ripen on the *interior* of sacs of various shapes called **asci**, and are freed by a simple tearing, disintegration, or opening of the ascus. In addition to the morels and truffles already mentioned, the *Peziza*, *Gyromitra*, and *Xylaria* are also part of this group. The asci usually contain eight spores.

Understanding Spore Formation

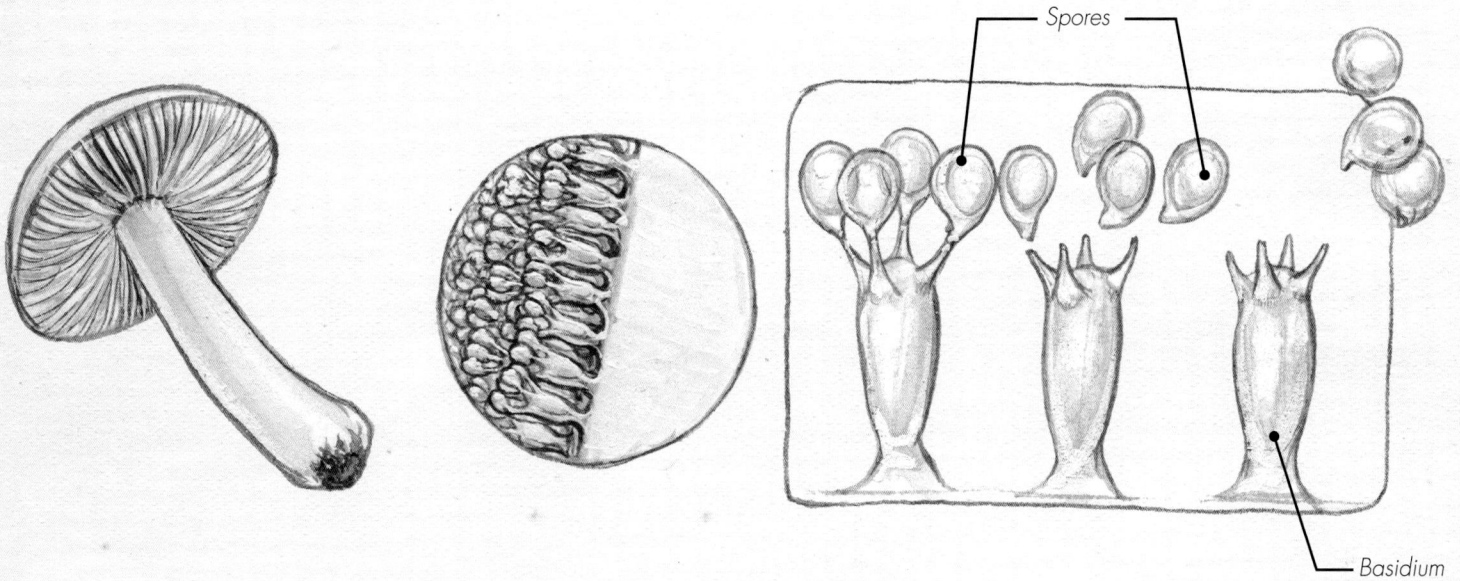

Spores

Basidium

The basidiomycetes

The spores are formed on the outside of cells called basidia.

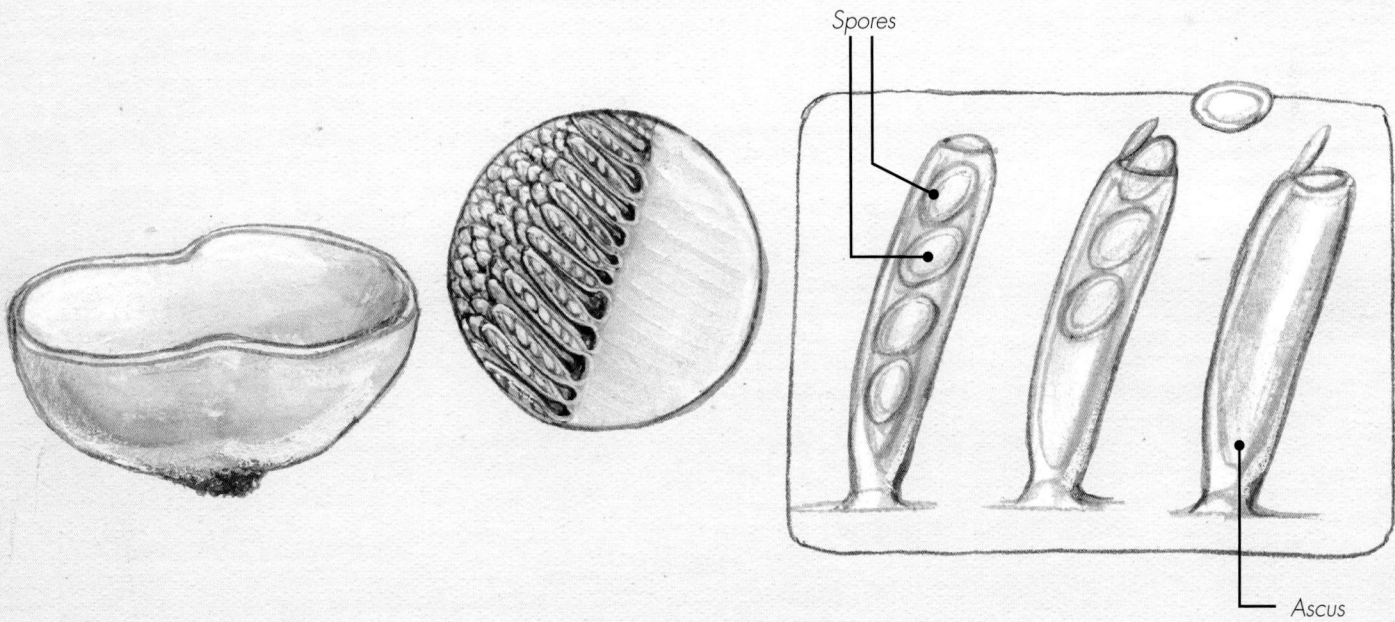

Spores

Ascus

The ascomycetes

The spores are inside sacs called asci.

Mushroom
Anatomy

Cap —

Gills—together, the gills form the hymenophore

Ring (partial veil) —

Stem or Stipe —

Volva (universal veil) —

▲ *Death cap (Amanita phalloides)*

Different Cap Appearances

▲ *Fibrillose*

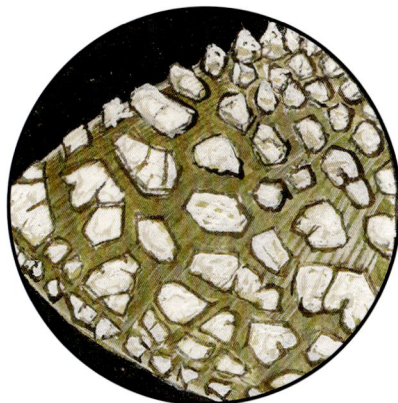

▲ *Warty*

▲ *Scaly*

Mushrooms come in such a variety of forms that trying to give an exhaustive description of their anatomy is essentially impossible. Over the course of their long evolution—it is estimated that they appeared on Earth 1.3 billion years ago—these organisms have acquired just about every shape imaginable. Even though the most well-known of these shapes is characterized by a cap set on top of a stem, that form alone has several variations, from the traditional silhouette of ceps to the less classically shaped morels, and all of the chanterelles and pieds-de-mouton (hedgehogs) in between. The almost limitless inventory doesn't stop there: we could also mention the *Peziza* cups, the clumps of certain *Clavaria*, *Xylaria* species with clubs, and more.

The major parts of a mushroom

Let's stop and look at an *Amanita* in detail, *Amanita phalloides* (p. 229). In essence, it consists of a **stem** (what mycologists sometimes call a "stipe") that supports a **cap**, creating a silhouette that resembles something like an umbrella. The inferior face of the cap is lined with radiating gills arranged in decreasing length: **lamella** and **lamellulae**. This gilled ensemble forms the **hymenophore**, the fertile part of the cap, which is carpeted by a fine layer of cells, the **hymenium**, that will produce **spores**, the mushroom's reproductive cells. The hymenophore is not necessarily gilled, however: it may also be made up of tubes, in the case of boletes, or teeth, if the mushroom is a member of the *Hydnum* family. It may also be smooth or wrinkled as we observe in black trumpets and chanterelles.

If we move down the stem of this *Amanita*, we will encounter a sort of ruffled skirt called a **ring**: mycologists call this a **partial veil** which, in certain mushrooms, protects the hymenophore in specimens that are not yet mature.

The cortina, ▶ another example of a partial veil.

Different Ring Appearances

▲ Descending "skirt" ring

▲ Ascending ring

▲ Double ring

This veil may be membranous, as it is in mushrooms with a well-developed ring, or delicate and made of fine filaments resembling a spider's web. In the latter case the ring is referred to as a **cortina** (drawing on p. 19), which we observe in the *Cortinarius* family.

Let's continue our journey down the stem of our *Amanita*: at the tip of its base, it dives into a kind of white filmy sac, the **volva** or **universal veil**. In order to understand what the universal veil is, it's best to imagine that the young *Amanita* is like an egg, and its shell is made up of this veil. As it grows, this "shell" tears open at the top and allows the mushroom to grow up and out while still maintaining its original position at the base of the stem. In other *Amanita*, like *Amanita muscaria*, the volva is much less membranous and considerably more fragile: as it blooms, the mushroom will tear it in multiple places,

forming many small pieces of volva that, in the adult mushroom, will be scattered at the bottom of the stem and on top of the cap, creating its characteristic "geography map" patches (see drawing on p. 19 and diagram below). Certain *Amanita*, like those belonging to the *Amanita vaginata* group, have only a universal veil and no partial veil. This is why they usually display a well-formed volva even though they never have a ring.

An important feature: lamellae

The way the hymenophore attaches to the stem is a very important feature for mushroom identification and is usually easy to observe. Based on this one criterion, mushrooms are broadly distributed into four large groups (for the sake of simplicity, we will describe the case of the gilled hymenophore, but the same terms can be used in all other cases, as well).

Different Volva Appearances

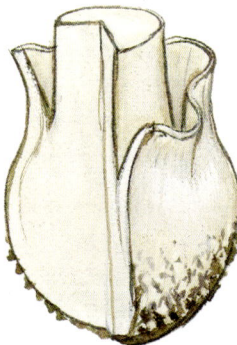

▲ Membranous volva

▲ Volva with ridges

▲ Circumcised volva

- The gills make no contact with the stem: in this case we are talking about **free gills**. Free gills are found in *Amanita* mushrooms but also in *Lepiota* and *Volvariella*; an extreme case can be seen in certain large *Lepiota* (genus *Macrolepiota*) whose gills terminate so far away from the stem that they create a smooth ring-shaped zone around the top of the stem, sometimes with a small ridge. These gills are described as being **collared**, and the smooth ridge is called a **collarium**.

- The entire width of each gill is in contact with the stem: these are **adnate** gills.
- The gills make only partial contact with the stem, often displaying a small indentation just before the stem: these are **emarginate** gills and are typical among *Tricholoma* and *Cortinarius*.
- The entire width of each gill is in contact with the stem and the gills extend down onto it in a net formation: these are **decurrent** gills and are characteristic of *Clitocybe* spe-

cies, among others, though not all *Clitocybes* necessarily have decurrent gills. We might also mention here the clearly decurrent folds of the chanterelles.

Naturally, there are many intermediate cases among these four groups, but learning to recognize these major distinctions will be very useful to you.

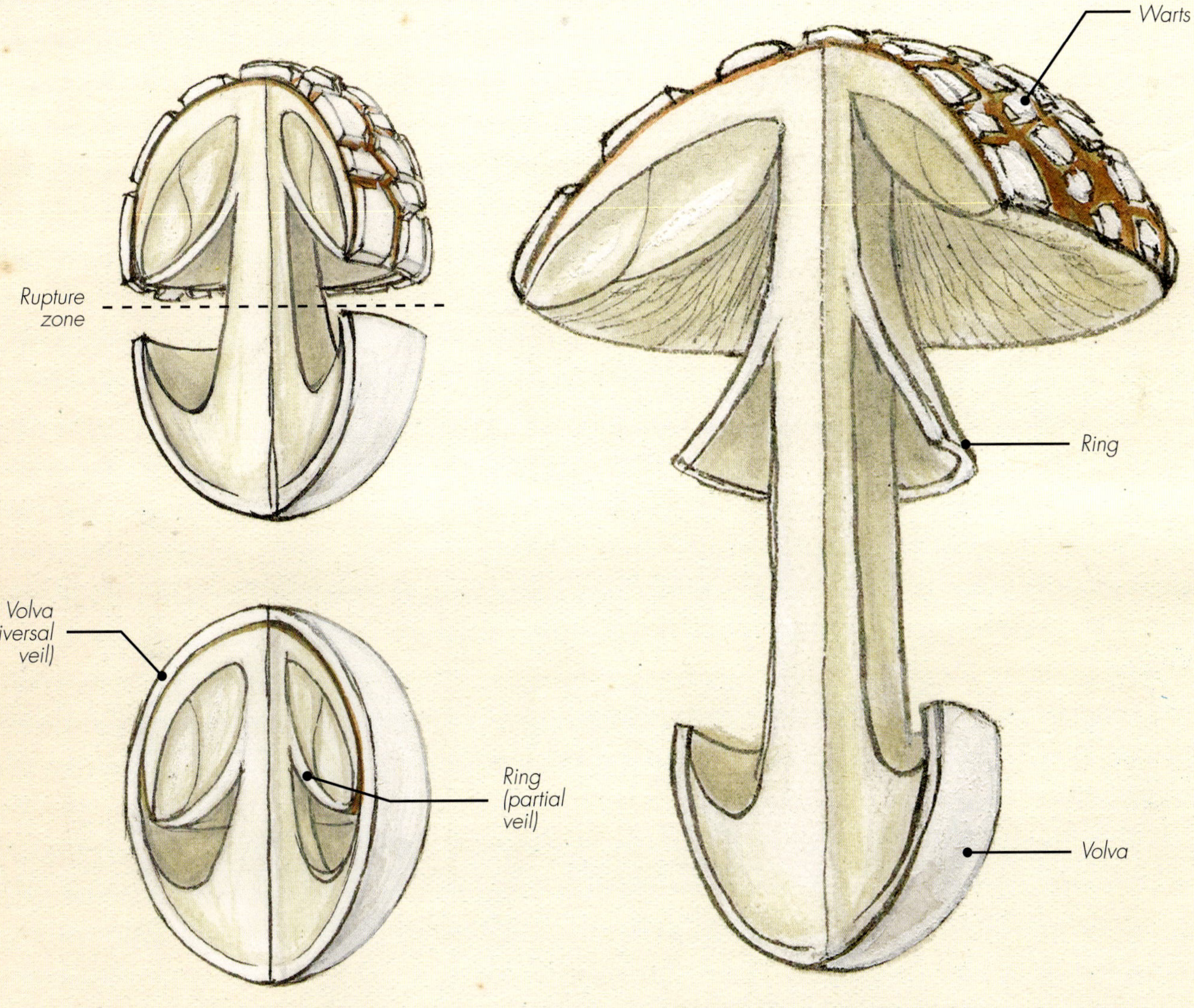

Rupture zone

Warts

Ring

Volva (universal veil)

Ring (partial veil)

Volva

Stages in the development of an *Amanita*

▼ Different Cap Shapes

Convex Conic Bell-shaped Umbonate Funnel

▼ Gill Attachments

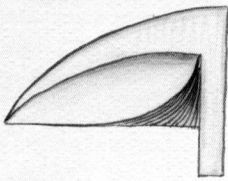 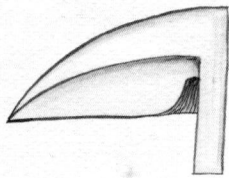 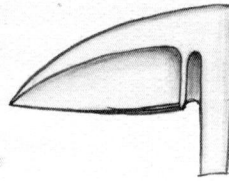 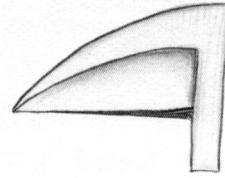 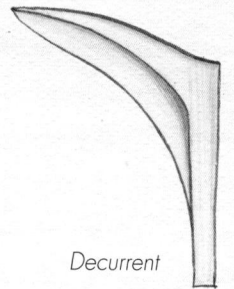

Free Emarginate Collared Adnate Decurrent

▼ Different Types of Hymenophores

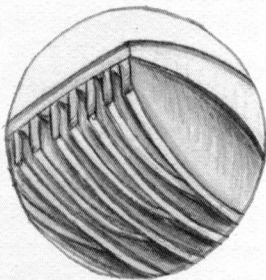 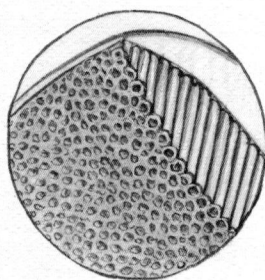 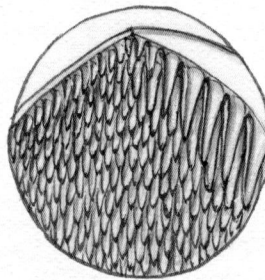 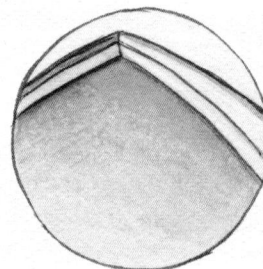 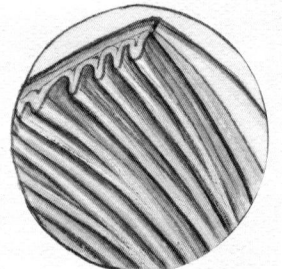

Gills Tubes Teeth Smooth Folds

▼ Different Stem Base Shapes

 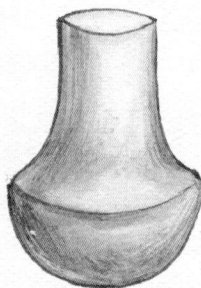    

Bulbous Marginate Bulb Cylindrical Fusoid Radicating Ventricose

▲ Pleurotus: *the cap is funnel-shaped, the gills are decurrent, and the hymenophore has folds.*

Identifying
Mushrooms

Identifying mushrooms is not easy: it requires not only knowledge but also, and perhaps most of all, a great deal of experience alongside expert mycologists. When you are starting out, it is very important to remember the fundamental principles we mention here; only by following them to the letter will you be able to deepen your understanding of the vast world of mycology and, more importantly, avoid making a mistake. In order to identify a mushroom with certainty:

• Only harvest mushrooms that are in perfect condition and, if possible,

at every stage in their development. Certain mushrooms change dramatically between their young stage and their adult stage. If you do not pay attention to these changes, you will not know how to recognize them.

• Scrupulously record the habitat of the mushrooms you are harvesting. Is your mushroom growing in the grass? Are there trees nearby and, if so, which ones? Is it growing in a deciduous or coniferous forest? Is it coming directly from the open ground or is it connected to dead wood? A few basic principles of botany will be quite useful

for identifying the trees with which a mushroom may be associating if the mushroom is mycorrhizal (see Symbiotics p. 29).

• Using a knife, gently lift up the base of the stem of the mushroom you want to identify. If you are not careful, you risk leaving part of a root, volva, or other important identification feature in the ground. Of course, if you are only harvesting mushrooms you already are familiar with, it is better to cut the base of their stem to avoid removing too much soil. They will also be easier to clean.

▲ Delicately lift the base of
the stem

▲ Dig up the mushroom
using a knife

• Place your mushrooms in a basket with a flat bottom. Avoid gathering in plastic bags because specimens can get mixed together and become damaged. If possible, place small containers in the bottom of your basket (cut off the ends of plastic bottles, for example) to separate each one of your discoveries. Mycologists often use plastic tool boxes or waxed paper for this purpose, which is quite practical.

• When you return home, it is very important to perform a **spore print**: to do this, separate a cap from its stem and place it flat on a sheet of paper or, even better, a sheet of transparent plastic (the transparency sheets used with overhead projectors are ideal when cut to the proper dimensions). If you cover the cap, necessary humidity will be maintained, and the spores will fall in a pattern resembling the hymenium of the mushroom. A few hours later, the spores propelled from the gills (or tubes, teeth, etc.) will be deposited in a dense mass on the sheet. You can then observe their color and avoid any unpleasant surprises. With a spore print, you will be able to distinguish the brownish-

pink spore print of the bitter bolete (p. 221) from the bright rust-colored spore prints of the *Cortinarius* and the brown of the agarics.

In any case, let's say it again:
• No identification guide is exhaustive. If your mushroom does not possess all of the characteristics indicated in your book, it is probably a different species and may be toxic.

• Never consume an unknown mushroom without showing it to a mycologist first, identifying it as edible. Mycological societies often have information available on their websites pertaining to color, shape, location, and edibility of mushrooms in their region.
• Never eat a mushroom in large quantities or over several consecutive meals.

▲ *The spore print allows you to study the colors of the spores, another indicator for identification*

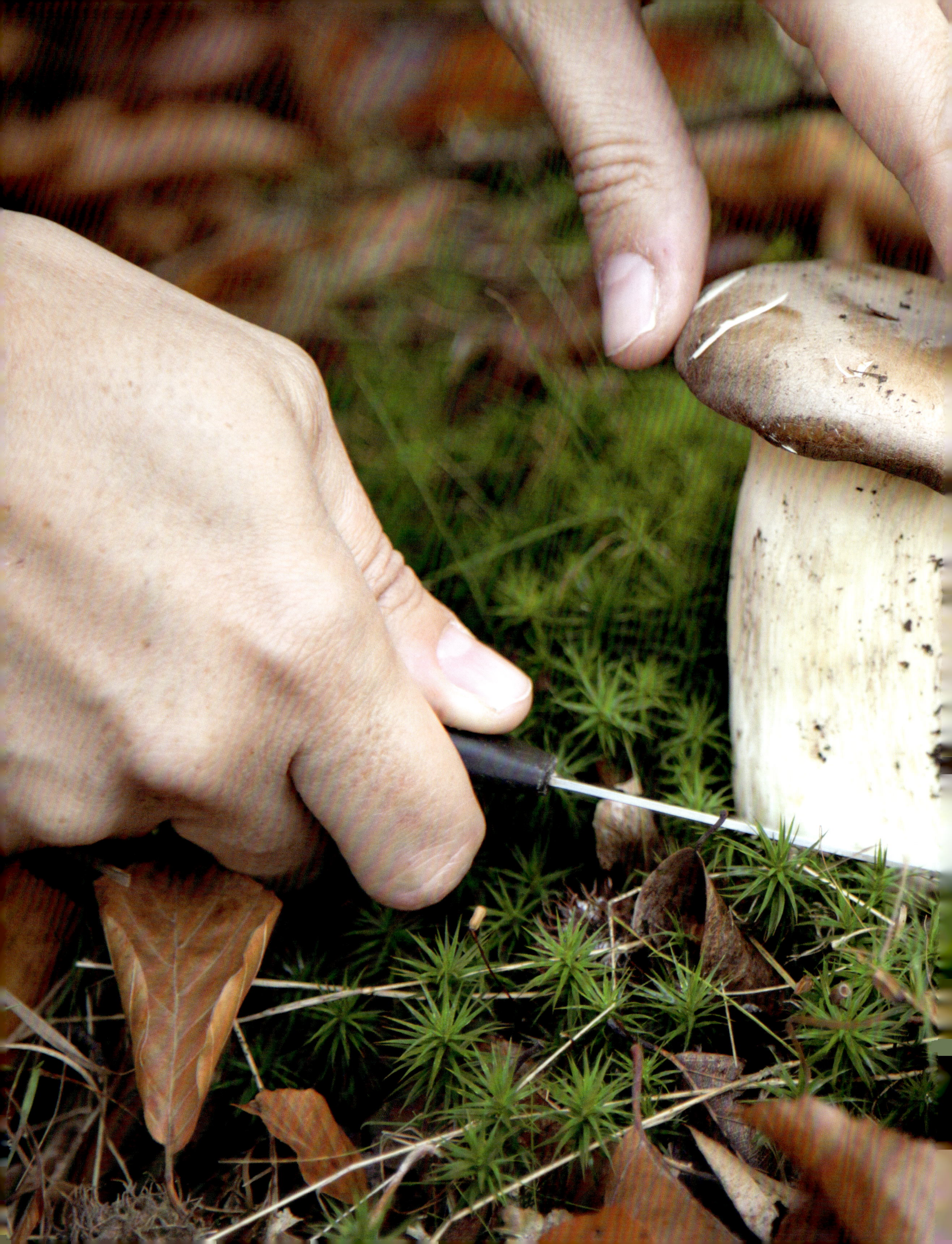

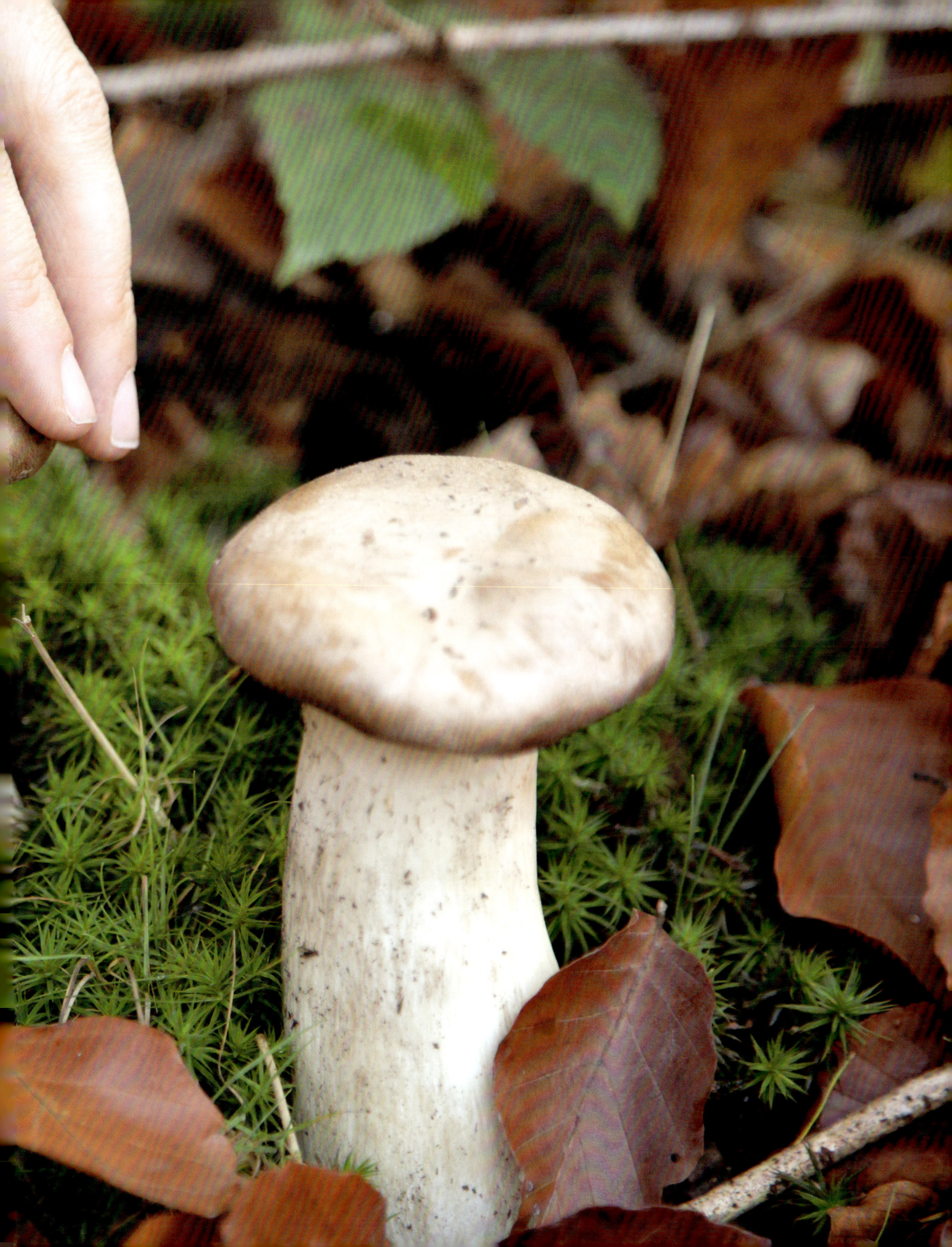

The Life
of a Mushroom

Mushrooms are divided into three groups as a function of how they feed themselves:

• Many mushrooms draw the substances they need to grow from other organisms that are either dead or decomposing: these are the **saprotrophs** (from the Greek *sapros*, "rotten, putrid" and *troph*, "food or nourishment").

• Other more aggressive mushrooms live at the expense of living organisms that they often weaken and can even kill: these are the **parasites**.

• The last group is much more interesting from a biological point of view, and significantly more diverse: this group brings together all of the mushrooms that have, over the course of evolution, formed mutually beneficial exchanges with other living things known as **symbiotic** relationships.

The saprotrophs

Mushrooms that feed on decomposing organic matter play a role in soil renewal and, along with bacteria, are the most important decomposers on the face of the earth. Without them, organic matter would accumulate year after year to astonishing heights. It is estimated that in a temperate forest, mushrooms decompose around 100 tons of organic matter per hectare (roughly forty tons per acre) per year. Saprotrophic mushrooms are everywhere and are sometimes highly specialized: some decompose wood by attacking each of the long molecules it is made up of; others go after leaves and other non-timber plant products; and still others break down hair, feathers, and other animal skin products, eventually followed by their bones.

Baker's yeast (*Saccharomyces cerevisiae*), a saprotrophic monocellular ascomycete, has the ability to grow without oxygen by releasing carbon dioxide, which is very useful to those of us trying to get bread dough to rise or produce beer and other fermented beverages. On the other hand, when saprotrophic mushrooms attack ancient manuscripts or invade certain dark and humid places that humans would prefer to keep for themselves, it can be nearly impossible to eradicate them, causing headaches for many curators. The mushrooms growing in colonies on the walls of the Lascaux caves in Dordogne are a well-known and shocking example: even though they are not attacking the cave paintings themselves, they mask them with abundant mycelia, endangering the 17,000-year-old drawings.

The parasites

Parasitic mushrooms take part in generation renewal by attacking weakened organisms and are for this reason called "weak" parasites, like many polypores. A few of these parasites create fungal infections that are sometimes difficult to get rid of, and many of us have fallen prey to them at one time or another. You are probably familiar with the onychomycosis fungi, for example, which deforms and nibbles down our nails, as well as dermatomycosis, which discolors our skin. While only mildly irritating if we are in good health, they can spread uncontrollably in individuals who are weakened or have failing immune systems. All plants have their own fungal parasites, often very specific ones at that, but animals suffer from fungal infestations, too. *Batra-chochytrium dendrobatidis*, for instance, is a lethal fungal parasite that attacks certain amphibians (frogs). It is currently monitored on a global scale because it is held responsible for massive loss of life among certain populations.

The mildews—parasites well-known to many plants, particularly those that are cultivated (grapevine downy mildew, *Plasmopara viticola*; potato blight, *Phytophthora infestans*, etc.)—are no longer considered to be fungi, and today are grouped alongside certain brown algae. Their spores have two flagella and must use a thin net of water in or-

der to move around, which explains why the development of the diseases they provoke is intimately related to atmospheric conditions.

The symbiotics

Biologists have a habit of saying that mushrooms are the "kings of symbiosis," and this reputation is holding strong. A **symbiotic relationship** is an association with mutual benefits between two or several living beings. Throughout geologic time, mushrooms have constructed relationships with a great number of living things, plants and animals alike. One of the most well-known relation-

ships is the one that exists between a mushroom and an alga or cyano-bacteria (a kind of bacteria able to perform photosynthesis): the organisms that result from this relationship are known as **lichens**.

Lichens are everywhere, from the summits of mountains to the edge of every sea and ocean; from the arid regions of deserts to the deepest equatorial rainforest and car bodies in the junkyard. With the great advantage of their "mixed" anatomy and the intimate connections between the two parties, they are capable of settling and thriving just about anywhere. Certain lichens, however, are sensitive to atmospheric pollution,

▼ *A mature Mycena species on a log...*

▲ Various saprotrophic mushrooms

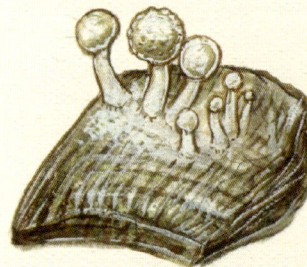

▲ Saprotrophic mushrooms developing on bones and human nails

▲ Various soil and dead wood ascomycetes

which makes them precious allies when it comes time to evaluate the air quality in a given area.

Mushrooms also associate with plant roots to form **mycorrhizae**. Almost all plants on Earth—more than eighty percent of species—possess mycorrhizae and are therefore **mycorrhizal**. This relationship is a very ancient one, and mycorrhizae have been identified in the roots of plant fossils dating back 400 million years. Biologists are even convinced that it is this relationship that allowed plants to colonize dry land. There is good reason to think, then, that if the mycorrhizae had not appeared, the earth would have never been covered with vegetation, or at the very least it would not at all have the appearance we are used to today.

▼ Batrachochytrium dendrobatidis *is a fearsome amphibian parasite*

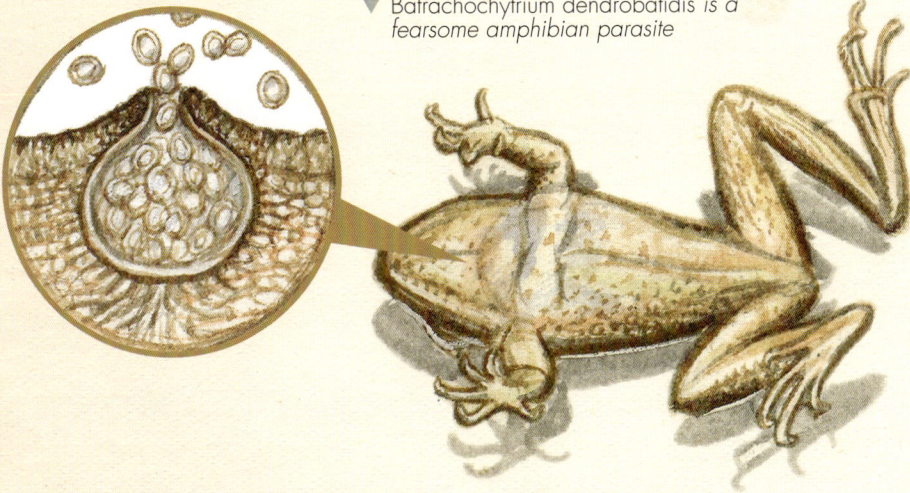

▼ Rhytisma acerinum, *a very common parasite on maple leaves*

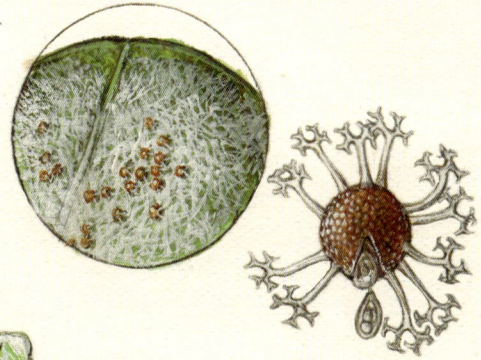

▲ *The oak tree powdery mildew forms a white powder on its leaves*

▼ *Molds are microscopic mushrooms that can be useful or damaging, depending on the species*

▲ *Parasitic mushrooms*

▲ *Chicken mushroom or sulfur shelf* (Laetiporus sulphureus)

Yeasts are unicellular mushrooms

The algae's green cells

A mushroom-like reproductive organ

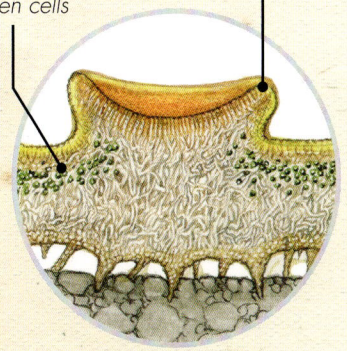

Like all lichens, Xanthoria parietina is formed by an association between the fungi mycelium and algae cells

The mycelium of mycorrhizal mushrooms associates with fine tree roots to form mycorrhizae

The Importance of
Spore Dispersal

The ability to form millions, even billions, of spores is critical to the life of mushrooms because, like all other organisms on earth, their survival depends on their reproductive potential. For millions of years, the mushrooms that were able to perform the phenomenon of spore dispersal most effectively were those that evolution selected. We can point to the *Cyathus*, which eject their peridioles with the help of a small spring, ascomycetes like the *Peziza* and the morels that use a kind of "spore cannon" called an ascus (p. 16), or the *Pilobolus* (see this page), miniscule and elegant mushrooms that grow on cow manure and the dung of other animals and propel batches of spores using a kind of balloon that suddenly deflates. But the phenomenon known as the "ballistospore" is perhaps by far the most astonishing of all.

As we have seen, it is easy and extremely useful to perform a spore print (see p. 25): the mature spores

"Sac" containing the spores. Propelled up to two yards, it will stick to the grass around it and be consumed by herbivore animals.

Vesicle propelling the "sac" of spores

▲ The strange and tiny Pilobolus

liberated by the mushroom are deposited densely, allowing us to see their color. But what is actually happening when these spores detach from the cells that conceived them? Is it a passive phenomenon in which the spores fall the way an apple falls from its tree? Or are they projected? The first hypothesis was favored at one time, but certain clues—like the abundant spore print from the cap of certain *Ganoderma*, for example—pointed to a phenomenon that was more complex. In reality, we have to distinguish between two groups of mushrooms: those that liberate their spores in a passive manner, and those that project them. Some mushrooms like puffballs (pp. 273, 275, and 276), truffles (pp. 155, 157, and 159), and other subterranean mushrooms with a tuber shape form their spores in an enclosed space, a kind of "stomach" in which any projection of spores would be futile. Instead, the spores are liberated passively by detaching themselves from their basidium (in basidiomycetes) or by the destruction of the ascus that holds them (in ascomycetes): these are **statismospores**. Many of the subterranean mushrooms in this group are consumed by animals (from earthworms to large mammals), which suggests that spore dispersal takes place after passage through the digestive system of these "transporters." Spores are also liberated passively in mushrooms from the *Phallus impudicus* group (p. 299). In this group, the spores are stuck inside a dark olive-colored and strong-smelling substance, the gleba, which attracts numerous insects. These insects land on this attractive viscous envelope, sometimes ingesting part of it, and then leave with spores sticking to their legs or in their digestive system, ready to be released elsewhere.

We know of 30,000 other mushroom species, however, that use an entirely different process to get rid of their reproductive cells. All mushrooms with folds, gills, tubes, or teeth—in other words, the large majority of basidiomycetes—eject their spores so they can be taken by the wind and transported as far as possible: these are **ballistospores**. This ejection requires a physical phenomenon that is fairly complex and only partially understood. In basidiomycetes, spores are formed on top of sterigmata, rigid extensions of the basidium. With the help of very sophisticated cameras, researchers observed a drop of water forming at the base of the spore. This drop, called "Buller's drop" in honor of the biologist who discovered it, forms as a result of the accumulation of sugary hydrophilic substances at a specific location on the spore. This drop is relatively heavy and displaces the spore's center of mass on top of its sterigma. When the drop reaches a critical size, it will come into tangential contact with the spore's surface, creating a "surface tension catapult" mechanism. This sharp displacement of weight will violently liberate the spore from its sterigma and subject it to an acceleration that can be more than 10,000 G, or more than 3,300 times what astronauts feel during the liftoff of a space shuttle.

Storing
Mushrooms

You've just come home with a beautiful basket of correctly identified mushrooms. If you decide not to eat them right away, you're going to have to store them somehow. Now, the majority of mushrooms can't be saved for very long: the shaggy inkcap (p. 69), for example, must be cooked directly after it is harvested. If you simply have too many to eat all at once, save some of them in the refrigerator to eat the next day. Cut the mushrooms in half after cleaning them and store them with the sliced side down.

Certain species can be preserved in other ways:

Drying

This method only works for species with firm and fragrant flesh. Wash the mushrooms, cut them into thin slices, and lay them next to each other in a warm, dry place no warmer than 110°F, preferably on a cooling rack. You can also use a machine specially designed for drying fruit. As soon as the mushrooms are dry, put them in a jar or can sealed with a screw-on lid or an airtight plastic bag. Avoid any prolonged contact with humid air. If the mushrooms are very dry, they will last several years.

Powder

Dry mushrooms can be reduced to a powder that will bring a savory note to your cooked dishes, soups, and sauces. The powder must be cooked at the same time. It can be saved for about one year.

Freezing

Cut the mushrooms into halves or quarters and blanch them in salted boiling water. Cool them in ice water before placing them in suitable containers. Place them in the freezer (rapid freeze) where they can be kept for three to six months. You can also freeze mushrooms that have already been cooked.

Canning

It is also possible to preserve mushrooms by pickling and canning them. Each specimen just needs to be blanched in boiling water first (timing varies depending on the species). The saffron milkcap and the bleeding milkcap go well with white wine vinegar, and you can also add honey, cloves, bay leaves, shallots, and/or garlic.

▲ *Black truffle (Tuber melanosporum)*

Buying
Mushrooms

It is important to note the differences between cultivated and wild mushrooms. Today we know how to cultivate around 100 kinds of mushrooms, only thirty of which are farmed industrially. They are all saprotrophs (see p. 28). Unfortunately, most of the very good edible mushrooms (chanterelle, truffle, etc.) are mycorrhizals that we are not able to cultivate, and all of those sold in stores come from forest harvesting. In the case of the truffle, it is possible to plant them in suitable terrain near trees with roots that associate with the truffle mycelium, but the outcome of these plantings is still too inconsistent.

Mushrooms like the common button mushroom (p. 46), shiitake (p. 80), *Pleurotus* species (pp. 109–111), and the nameko (p. 92) are also available no matter the season. These mushrooms are not very difficult to grow, and their production has been so perfectly mastered that in all but rare cases they are sold in perfect condition.

The same cannot be said for forest mushrooms, though, which typically present two major issues (we won't mention the fact that entire regions are pillaged to supply certain markets, a fact that constitutes its own range of human and environmental problems):

• Accurate species identification is not a guarantee. As astonishing as it may seem, the forest mushrooms that arrive at market in Central Europe, North America, and the United Kingdom never undergo any kind of control, for the simple reason that there are no inspectors to perform such controls. So, as a result, we may find bitter boletes (p. 221) on the same trays as chanterelles, no doubt ruining a few family dinners. There have also been reports of *Lactarius salmonicolor* (p. 192) being sold as *Lactarius deliciosus* (saffron milkcaps) for the modest price of fifteen dollars a pound!

• The quality of the harvest often leaves much to be desired. Would you eat a steak that has been left for a month and a half in the refrigerator? The answer is surely no. And yet many mushroom lovers—particularly around the holiday season—will readily buy moldy mushrooms in the supermarket even though the woods have been deserted for several weeks. It is difficult to know how many tons of mushrooms in poor condition—after being attacked by bacteria or other fungi, larvae, or frozen by the first frost—are sold each year and how many poisonings those unscrupulous vendors are responsible for.

If you are buying mushrooms in the market, you need to use the same purchasing criteria you would for any other perishable product:

• Only buy forest mushrooms in season, essentially in autumn, "when you know they are growing," as the sacred saying goes.

• Choose vendors known for being responsible and buy mushrooms that you can identify yourself. Do not trust vendors who offer you a taste of a mushroom you are not familiar with.

• Finally, only select mushrooms that are in perfect condition. Take the time to examine them in order to detect traces of mold or softening, which is often a sign that they are beginning to decay.

Mushroom
Poisoning

Whether you are a very careful mushroom consumer or a person who regularly tests new species, never forget that mushrooms are responsible for serious poisonings every year. There are two kinds of poisonings, depending on the length of time it takes for the first symptoms to appear. Delayed appearance of the first symptoms of poisoning is often an indicator of severity.

Early Onset Toxicity (appearance of symptoms within less than six hours)

Gastrointestinal syndrome

Between fifteen minutes and two hours after the ingestion of mushrooms. Frequent and less severe, this syndrome presents as a mild gastroenteritis. The poisoning may be linked to the consumption of too many edible species, consumption of species rich in laxative substances (*Suillus*, *Ramaria*, etc.), or a sensitivity to certain species like the clouded agaric (p. 61).

Resinoid syndrome

Up to five to eight hours. Nausea and vomiting accompanied by abdominal pain that may persist for some time. The tiger tricholoma (p. 209), the livid pinkgill (p. 187), the jack-o'-lantern mushroom (p. 199), the agarics in the yellow-staining group (p. 167), and red-pored boletes in the devil's bolete group (p. 219) are the mushrooms largely responsible for this poisoning.

Muscarinic or Sudorian syndrome

From fifteen minutes to two hours. Excessive perspiration, shrinking pupils, and gastrointestinal symptoms. Most often associated with the white *Clitocybes* (p. 181)—when confused with the sweetbread (p. 62)—the deadly fibrecap (p. 241), and the *Mycena* species in the lilac bonnet group (p. 197). Rarely deadly, it can nevertheless be very serious.

Pantherina poisoning

Between thirty minutes and three hours after ingestion. Digestive problems, hallucinations, mental confusion, and sleepiness. The fly agaric (p. 173), jonquil amanita (p. 170), and panther cap (p. 175, which contains the most poison) are the most commonly cited mushrooms related to this poisoning, which is rarely deadly but sometimes relatively serious.

Narcotine or psilocybin toxicity

From thirty minutes to two hours. Euphoria, loss of spatial awareness, hallucinations, anxiety, panic, and even acts of violence, which may be accompanied by convulsions and a potentially life-threatening coma if a strong dose is consumed. Species responsible: the mushrooms of the liberty cap (*Psilocybe semilanceata*) group (p. 246).

Coprine syndrome or the "Antabuse effect"

From thirty minutes to two hours. Severe nausea that can last several days, reddening of the face, excessive perspiration, and heartbeat irregularity. These effects do not appear unless there is a simultaneous consumption of alcohol. Species responsible: *Coprinus* specimens from the common inkcap group (p. 182).

Hemolytic syndrome

Less than six hours. Destruction of red blood cells, digestive problems, and nausea. Involves all species that have not been properly cooked: numerous *Amanita* (like the blusher, p. 53), certain boletes, and various ascomycetes like morels (pp. 145–146). May be serious in the event of high consumption.

Paxillus syndrome

Less than six hours. Acute intravascular hemolysis, probably due to an allergic phenomenon that can be very serious and even deadly. Species concerned: Brown rollrim (p. 245) and neighboring species.

Late Onset Toxicity (appearance of symptoms after six or more hours)

Phalloides syndrome

Between six and twenty-four hours after ingestion. Violent digestive symptoms and acute hepatitis that may become sudden and severe between the third and fifth day. Species responsible: the death cap (p. 229) and related species, the mushrooms of the funeral bell group (p. 237), and the *Lepiota* of the deadly dapperling group (p. 243). This very serious poisoning is responsible for the majority of deaths caused by mushrooms.

Orellanine syndrome

Between twenty-four and thirty-six hours after ingestion. Digestive problems and intense thirst. Followed by impaired renal function between the fourth and fourteenth day, with acute renal failure that may become chronic. Species re-

sponsible: various *Cortinarius*, but most of all the fool's webcap group (p. 233).

Gyromitrin syndrome

Between six and twelve hours. Digestive problems and headache with fever. May be followed by severe liver and renal failure (from thirty-six to forty-eight hours after) along with mental confusion. This type of poisoning may be deadly when the hepatitis is sudden and severe. Species responsible: various ascomycetes, the false morel being the species most often blamed (p. 248).

Proxima syndrome

Between eight and fourteen hours. Digestive problems followed by moderate hepatitis and renal failure. Principal species responsible in Europe: *Amanita proxima* (p. 171). In North America: *Amanita smithiana*. Full recovery is usually possible after renal dialysis.

Rhabdomyolysis syndrome

Between one and three days. General fatigue and muscle pain accompanied by significant perspiration. Muscular rigidity begins one or two days later and can complicate into cardiac arrhythmia leading to the death of the person

who has been poisoned. It usually occurs after a high consumption of *Tricholoma* in the yellow knight group (p. 247).

Acromelalgia syndrome

Up to twenty-four hours. Very painful burns at the extremities of the hands and feet, which may last several months. Principal species responsible in Europe: paralysis funnel (p. 179). Similar species in North America: *Clitocybe inversa*, *squamosa*, *gibba*, and *Hygrophoropsis aurantiaca*.

Flagellate dermatitis

Between one and three days. Allergy to shiitake (p. 80) observed several times in Europe, North America, and well-known in Asia. A cutaneous eruption with a whip-like appearance that can be more or less itchy is the major symptom.

Szechwan syndrome

Frequent risk of hemorrhage resulting in platelet changes and most notably the presence of purpura (cutaneous hemorrhage). It is caused by excessive consumption of Judas ears (p. 152) or neighboring species (the "black mushrooms" in Asian cuisine).

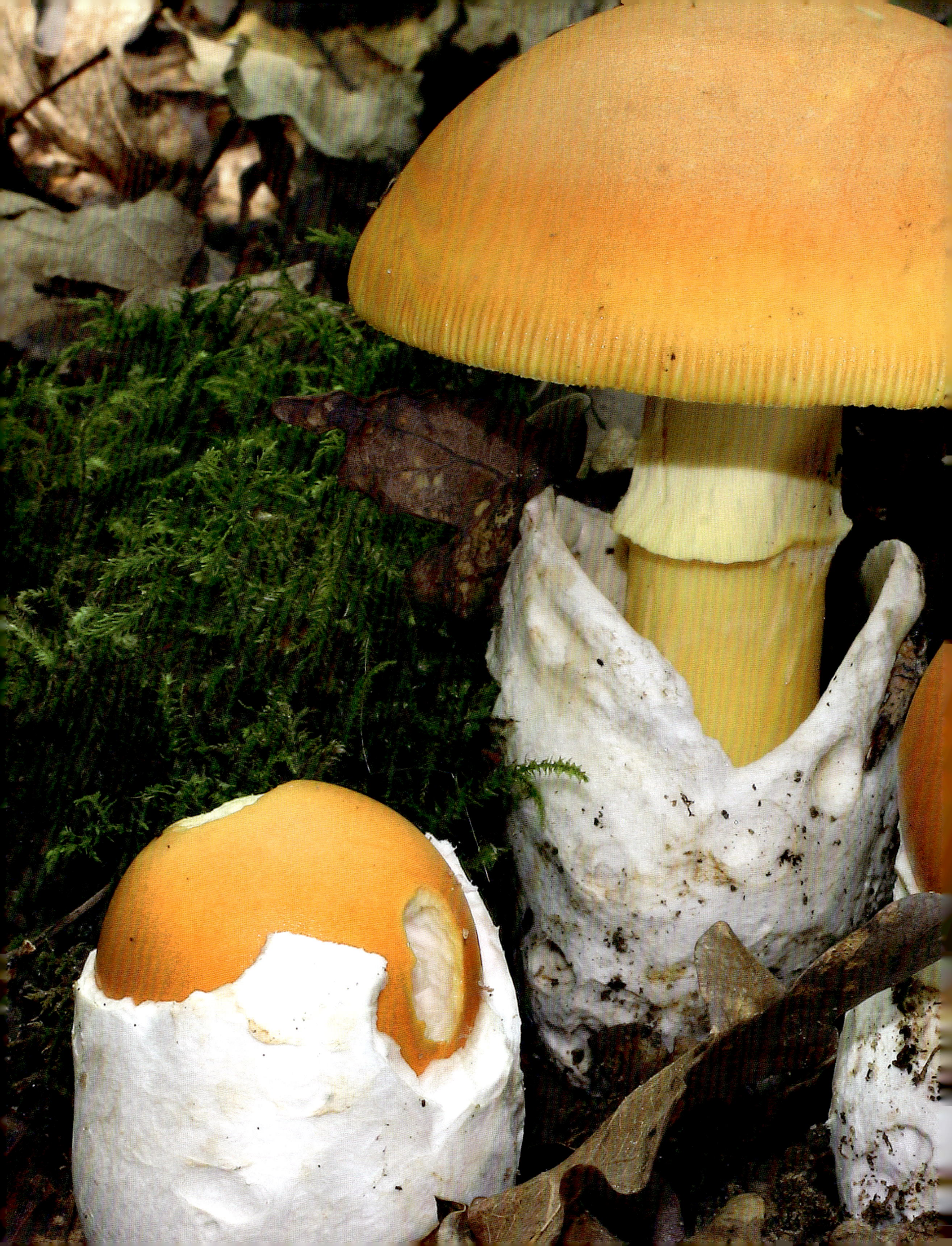

Edible Mushrooms

- ◉ **Cap:** *up to 4 inches (10 cm) wide, white progressing to pinkish-gray as it ages, smooth or scaly.*
- ◉ **Gills:** *free and crowded, bright pink then brown; spore print chocolate brown.*
- ◉ **Stem:** *white, typically tapered at the base, with a small and delicate ring that tends to disappear in maturity.*
- ◉ **Flesh:** *firm; sweet flavor and pleasant mushroom odor.*
- ◉ **Habitat:** *grass in lawns, meadows, and pastures.*
- ◉ **When to harvest:** *All year long, apart from periods of severe cold, but most of all in summer and autumn.*

▲ Gill maturation in the meadow mushroom (Agaricus campestris): *note the change in color.*

Meadow mushroom

Agaricus campestris (Fr. : Fr.) Link

The meadow mushroom is probably the most well-known and most common of the wild agarics. Even though its close cousin the common button mushroom (p. 46) might steal first place among agarics, as a whole it only arrives on our plates in its cultivated form, with a white coloring selected by farmers for many years that is significantly different from its wild brown color. The meadow mushroom can usually be seen from afar in the fields and pastures it loves because it often forms fairy rings of small white caps.

In the kitchen

Meadow mushrooms and button mushrooms are excellent edibles that can be eaten raw, which is very rare for mushrooms. Eat them with a pinch of salt or in a salad, only adding seasoning at the last minute so as not to oversoften this mushroom's pleasantly firm flesh.

Warning

Harvesting meadow mushrooms can be an enjoyable activity either alone or with family, but it is important to be vigilant and not let yourself be fooled by a toxic lookalike.

The white *Amanita* (pp. 229–231) are deadly and always grow near the trees they are associated with. Their gills are white, their stem bears a distinct skirt-shaped ring, and, most importantly, they emerge from a membranous sac volva that surrounds the base of the stem. When harvesting mushrooms, take care not to overlook a feature as important as a volva.

The yellow stainer (p. 167), which is difficult to digest and also plumper, is considered toxic, gives off an unpleasant inky odor, and yellows when exposed to air after being cut, especially at the bulbous base of its stem. See also *Agaricus bresadolanus* (p. 164).

The frequent confusion between the meadow mushroom and the white dapperling (p. 85), which grows in the same areas, is harmless because the latter is not poisonous: it can be distinguished by its white or pale pink gills—never brown—and its club-shaped stem with a well-formed ring.

Fairy ring

Certain mushrooms have a tendency to grow in circles, but there is no need to reference magical or extraterrestrial powers to explain these structures. They are the result of a natural developmental phenomenon: mushroom spores, if the conditions are favorable, will germinate in a very specific location. The mycelium that arises from the spores grows in every direction and, if it could, it would form a perfect sphere. Its development, however, is limited in depth by the lack of oxygen underground and in height by gravity in the open air. As a result, all that appears on the ground is the outline of a circle that may be more or less uniform in shape, depending on obstacles such as trees, roots, and irregularities in the terrain.

Common button mushroom

Agaricus bisporus (J. Lange) Imbach

- ◉ **Cap:** *up to 4 inches (10 cm) wide, brown in the wild, but most often white when sold commercially.*
- ◉ **Gills:** *free and crowded, bright pink then brown; spore print chocolate brown.*
- ◉ **Stem:** *white or brownish, cylindrical or club-shaped, with a large thick ring that is triangular in cross section.*
- ◉ **Flesh:** *firm; sweet flavor and pleasant mushroom odor.*
- ◉ **Habitat:** *in grass, but also under trees and sometimes in dunes.*
- ◉ **When to harvest:** *All year long, apart from periods of severe cold, but most of all in autumn.*

The button mushroom is a beautiful mushroom that is well-known and widely commercialized. The forms that grow in nature are brown, but the commercial button mushrooms are in general quite white: this is simply the result of selection by farmers who realized that the white strains sold better than the brown ones. Specialists are not yet in agreement about the Latin name for the button mushroom: some think that Charles Horton Peck, an American mycologist, had already described it with the name *Agaricus brunnescens* some twenty-six years before Jakob Emanuel Lange proposed the name *Agaricus bisporus*. The older name should logically be the one that is kept, so the American name is the one we should probably be using. *Agaricus campestris* and *A. bisporus* look so alike that sometimes they can only be differentiated through microscopic characteristics.

In the kitchen

The consumption of raw mushrooms is usually strongly discouraged. Button mushrooms seem to escape this rule, though: they are one of the rare mushrooms that can be eaten raw, much to the delight of salad lovers. They are sold abundantly in markets and large grocery stores. Choose ones that are young—the pink color on the gills, not purely brown, is a helpful clue—and firm without traces of browning.

Giant horse mushroom

Agaricus osecanus Pilát

- ⊙ **Cap:** *up to 8 inches (20 cm) wide, white then yellowing as it matures.*
- ⊙ **Gills:** *free and crowded, whitish at first then pinkish-gray, brown in maturity; spore print chocolate brown.*
- ⊙ **Stem:** *white, cylindrical or club-shaped, fluffy at the base, with a large ring whose underside is lined with triangular flakes.*
- ⊙ **Flesh:** *firm; sweet flavor and pleasant bitter almond odor.*
- ⊙ **Habitat:** *grass in fields and meadows.*
- ⊙ **When to harvest:** *Autumn.*

Even though this mushroom is not the largest of the agarics, it still reaches a respectable size that makes its white cap easy to spot from a distance in green grass. It is most often confused with *Agaricus macrocarpus*, which grows mainly in the woods. *Agaricus urinascens* (the macro mushroom) is also a white agaric, but it is a true giant: its cap can reach up to 12 inches (30 cm) wide. The distinction between these various white agarics is often a delicate matter, even for expert mycologists. Fortunately, they are all edible as long as they are harvested in areas free of pollution. Also watch out for the yellow stainer (p. 167), which is highly indigestible, toxic, and has flesh that turns yellow if broken, particularly at the base of the stem. And, as always when encountering white mushrooms, avoid deadly *Amanita*, which all possess a volva in a sac at the base of the stem.

In the kitchen

Given their size, the caps of the giant horse mushroom are easy to stuff if turned over and removed from the stem.

Blushing wood mushroom

Agaricus silvaticus Schaeff.

- ◎ **Cap:** *up to 6 inches (15 cm) wide, covered in reddish-brown to dull brown-gray scales on a dirty whitish background.*
- ◎ **Gills:** *free and crowded, pink-gray then dark brown with lighter edges; spore print chocolate brown.*
- ◎ **Stem:** *whitish, distinctly bulbous at the base with a ring that typically looks "dirty," brownish white; the surface of the stem, covered with incomplete wooly wreaths beneath the ring, turns bright red within minutes when scraped.*
- ◎ **Flesh:** *off-white turning red when cut; sweet flavor and pleasant odor.*
- ◎ **Habitat:** *in forests, especially under conifers.*
- ◎ **When to harvest:** *Autumn.*

As a general rule, the *Agaricus* are relatively difficult mushrooms to recognize, even for experts. With its affection for conifers, its brown scaly cap, its bulbous stem, and its habit of reddening, the blushing wood mushroom is an exception. There are, however, other agarics that share some of these characteristics: the scaly wood mushroom (*Agaricus langei*) has a scaly cap and flesh that turns red, but its stem does not have a bulb; the tufted wood mushroom (*Agaricus impudicus*) has darker scales that stand out against the cap's surface, and its flesh gives off an unpleasant rubbery odor. These last two are reported from the UK, but not from North America. The prince mushroom (*Agaricus augustus*, p. 162) is larger, its cap is covered with redder scales, and its flesh smells like bitter almonds.

Did you know?

Recent medical studies have shown that a diet including blushing wood mushrooms was beneficial to patients with certain kinds of colorectal cancers by lowering their arterial blood pressure and cholesterol.

▲ *Note the gills' change in color and the reddening of the flesh over the course of maturation.*

Wood mushroom

Agaricus sylvicola (vittad.) Peck

- ◉ **Cap:** up to 5 inches (12 cm) wide, smooth and silky, white and turns yellow when touched.
- ◉ **Gills:** free and crowded, pinkish white then pink, and finally brown; spore print chocolate brown.
- ◉ **Stem:** more or less cylindrical or a little club-shaped, whitish or rose, with a large white skirt ring.
- ◉ **Flesh:** white, yellowing slowly and slightly when cut; sweet flavor and strong aniseed odor.
- ◉ **Habitat:** in the forest, especially under deciduous trees.
- ◉ **When to harvest:** Autumn.

With its yellowing white cap, its slim and elegant appearance, and its pleasant scent of aniseed, the wood mushroom is easy to recognize. It is nevertheless possible to confuse it with its close cousin, the flat-bulb mushroom (*A. essettei*) (not reported from North America), which, as the name suggests, has a distinctly bulbous stem. This common confusion is fortunately not at all dangerous, but you must make sure not to inadvertently collect yellow stainers (p. 167), which are generally very difficult to digest, or toxic, and can be distinguished by their rather unpleasant iodine odor. As always with mushrooms that have gills and a ring, be vigilant when you harvest and examine the base of each stem carefully: if it emerges from a volval sac, you are in the presence of an *Amanita* that is potentially lethal. The *Volvariella* (p. 107) also have free and pink gills and their stem comes out of a membranous volva, but they do not have a ring.

Important

To avoid errors in mushroom identification, it is very important to keep in mind the color of a specimen's spore print (see p. 25 for how to make a spore print). The small cells are projected from the gills of mushrooms that possess them and are either colorless or of various shades. Those of agarics are brown, while those of the *Amanita* are perfectly white; another useful feature that will keep you from making regrettable mistakes.

- ⊙ **Cap:** *up to 8 inches (20 cm) wide, smooth, a beautiful orange or bright red-orange color.*
- ⊙ **Gills:** *free, crowded, beautiful yellow; spore print white.*
- ⊙ **Stem:** *yellow like the gills with a yellow ring, emerging from a large volva in a white sac.*
- ⊙ **Flesh:** *firm; sweet flavor and pleasant mushroom odor.*
- ⊙ **Habitat:** *under deciduous trees; most common in southern Europe.*
- ⊙ **When to harvest:** *Late summer and early autumn, particularly after the big summer storms.*

Caesar's mushroom

Amanita caesarea (Scop. : Fr.) Pers.

In the world

The Caesar's mushroom has given its name to an entire group of Amanita that are generally brightly colored and scattered just about everywhere around the globe. This group is named the "Caesareae section" by mycologists and today includes over fifty species, with even more that have yet to be discovered. Amanita basii, for example, is the Mexican Caesar's mushroom and is traditionally cooked in a stew with various boletes and aromatic herbs. Amanita jacksonii is the North American Caesar's mushroom, but is not reputed to have the same good flavor as the true Caesar's mushroom. Caesar's mushroom has yet to be reported from the UK, perhaps due to cool summer temperatures.

The reputation of the Caesar's mushroom, so named because it was a favorite of certain Roman emperors, is well-established. Not only is it a magnificent-looking mushroom, it also shares with the king bolete (p. 129) the honor of being considered the best edible mushroom. The Caesar's mushroom is harvested more and more even though it grows at a time of year when most gastronomes are not used to spending time in the forest. It associates with various deciduous trees, particularly oaks. Its marvelous red-orange cap, its egg yolk yellow gills, stem, and ring—very important characteristics to look for—along with its broad white volva make it an easily recognizable mushroom.

In the kitchen

The Caesar's mushroom is an excellent mushroom with fine and especially flavorful flesh that can be consumed raw, simply seasoned with a drizzle of olive oil, salt, and pepper. Certain recipes in traditional French-style cooking also describe stuffed orange caps but preparing them this way would cause the very unique taste of the mushroom's flesh to be partially masked by the filling.

Warning

The Caesar's mushroom is very easy to recognize as long as you can confirm that its gills, ring, and stem—which emerges from a large sac volva—are not white, but a deep yellow. Certain "faded" forms of the fly agaric (p. 173) bear a strong resemblance to it and may trick the amateur, but the gills, ring, and stem are perfectly white, and its very brittle volva is not membranous and tapers into a few flaky protrusions encircling the bulbous base of the stem.

Bearded amanita

Amanita ovoidea (Bull.) Link

- ◉ **Cap:** *up to 10 or even 12 inches (25 or 30 cm) wide, smooth, white or cream to pale beige.*
- ◉ **Gills:** *free, crowded, white or cream, with flaky edges; spore print white.*
- ◉ **Stem:** *white like the gills, emerging from a large white or russet membranous volva, bears a large white ring of a very creamy consistency (not at all membranous).*
- ◉ **Flesh:** *white; sweet flavor and a distinctive iodine, almost "from the sea" odor that becomes unpleasant when the mushroom matures.*
- ◉ **Habitat:** *under oak trees and on calcareous soil.*
- ◉ **When to harvest:** *End of summer and in autumn.*

Saying that this *Amanita* can be seen from afar is quite an understatement. Its imposing size and pale colors make it extremely difficult to confuse with many other mushrooms, provided that the harvester checks that its volva is white or almost white and that its ring, very poorly defined, has a consistency similar to that of whipped cream. However, as is often the case with mushrooms, this edible *Amanita* has a toxic lookalike, *Amanita proxima* (p. 171); this mushroom stands out because of its russet volva and, most of all, for its distinctly more membranous ring. It is also smaller and slenderer than its *ovoidea* cousin. Reports of *Amanita ovoidea* from the UK are rare, and it is not reported from North America.

In the kitchen

While it is consumed in certain regions, the bearded amanita is not very appetizing. Its flesh is rather soft, and the less than pleasant odor it releases makes it difficult to imagine what delicious dishes it could possibly be added to even after being cooked and seasoned. It seems that its consumption is for the most part anecdotal, and enthusiasts seem more attracted by the abundance of its flesh than by its quality.

Blusher
mushroom

Amanita rubescens Pers. : Fr.

- ⊙ **Cap:** *up to 6 inches (15 cm) wide, often with a "dirty" appearance, covered with irregular grayish warts on a background that is cream-colored, ochre-brown to reddish or pinkish-brown.*
- ⊙ **Gills:** *free, fairly crowded, whitish and stain quickly to a pink or reddish-brown color; spore print white.*
- ⊙ **Stem:** *bulbous or club-shaped, whitish under the gills, quickly becomes pink or reddish-brown going toward the base, with a ring that is white or has pinkish-brown markings.*
- ⊙ **Flesh:** *white; turns distinctly red where it is damaged; sweet flavor and mild odor.*
- ⊙ **Habitat:** *under both deciduous and coniferous trees.*
- ⊙ **When to harvest:** *Early summer to late autumn.*

The *Amanita* have a bad reputation, and rightly so; with so many species that are poisonous and even deadly, the edibles often go unnoticed. Nevertheless, the blusher is a truly good mushroom that happens to be very common throughout most of Central Europe and the United Kingdom. A very similar species in North America is sometimes called *Amanita amerirubescens*, because it is not quite the same as the European species. It can still be found in most field guides under *A. rubescens*. It is considered edible in North America, but with caution. In order to recognize and identify the blusher correctly, you must carefully examine the surface of the stem and locate the characteristic traces of reddening around bruises and insect bites: this reddening is obvious in older specimens, but it is often quite discreet in fresh young mushrooms. The greatest risk for error lies in the possibility of confusion with the panther cap, or false blusher (p. 175), which is highly toxic, much more elegant-looking, does not redden when bruised, and at the base of its stem is surrounded by distinctive white bulges. The gray spotted amanita (*Amanita spissa [excelsa]*) also resembles it a great deal, but its cap is browner, it does not redden, and its flesh gives off a turnip scent.

Important

The blusher, like morels, should not be consumed raw: it contains substances that are mildly toxic that are quickly destroyed by heat. For this reason, it must be cooked, and any water released during cooking should be discarded.

Grisette
mushroom

Amanita groupe *vaginata* (Bull. : Fr.) Lam.

- ◉ **Cap:** *up to 4 inches (10 cm) wide, gray, with striped and grooved edges.*
- ◉ **Gills:** *free, only slightly crowded, whitish; spore print white.*
- ◉ **Stem:** *slender, without a ring, emerging from a large, white, sac-like volva*
- ◉ **Flesh:** *white, thin; sweet flavor and mild odor.*
- ◉ **Habitat:** *under both deciduous and coniferous trees.*
- ◉ **When to harvest:** *Early summer to late autumn.*

Mycologists refer to all *Amanita* without a ring as "vaginate" *Amanita*. There are quite a few, and little is known about them, even now. Several species have been described in recent years, but few mycologists have the temerity to continue identifying them because the features that set them apart are so subtle and often difficult to observe. The fact remains, however, that in certain regions, vaginate *Amanita* with gray caps—while we cannot always name them individually—are edible as long as they are cooked. To cite a few common species, we will mention here *Amanita mairei*, which we often see beneath pine trees, and *Amanita simulans*, which is usually found under poplars, both reported only from Europe. The latter can be recognized by its rather imposing stature and its cap with an irregular surface that often looks like it has been hammered. As there are many unknown species of vaginate Amanitas in North America, they are not recommended as edible. The grisette is not reported from the UK.

Did you know?

All (or almost all) *Amanita* are mycorrhizal, meaning their underground mycelium associates with the roots of surrounding trees. Mycologists have realized in recent years that this obligatory association is also highly specific. Be especially attentive to the trees around you when harvesting vaginate *Amanita*, because they can provide vital information that will help you identify the *Amanita* associating with them.

Poplar
fieldcap

Agrocybe cylindrica (DC. : Fr.) Maire

- ◉ **Cap:** up to 4 inches (10 cm) wide, smooth and then veined or chipped as it ages, brown, beige, or white.
- ◉ **Gills:** emarginate, widely spaced, cream to brown; spore print brown.
- ◉ **Stem:** roughly cylindrical, fibrous, white, with a small ring that is brown on its superior face in mature mushrooms.
- ◉ **Flesh:** white, yellowing slowly and slightly when cut; sweet flavor and faint but pleasant odor.
- ◉ **Habitat:** in clumps on logs or living tree trunks, usually beneath poplar trees, but also on other deciduous trees, mainly in southerly regions.
- ◉ **When to harvest:** Autumn.

The poplar fieldcap, reported from the UK but not from North America, is far from being the most common of all the mushrooms that grow in clumps on wood, which is truly a shame because it is a very nice edible mushroom. It favors the warm areas in southern Europe, though it can also sometimes be found further north. Its cap is a blend of different brown tones that tends to crackle in mature mushrooms and when it reaches a large size. Its spores are brown and stain the gills and the top of the ring, where they sometimes fall accidentally as they mature. It is best not to eat the stem, which is very fibrous. The honey fungus (p. 58) on the other hand is very common and is sometimes confused with the poplar fieldcap. The scaly cap of the honey fungus and its gills, which are never brown and are somewhat decurrent, usually suffice to tell them apart.

Did you know?

Like the *Pleurotus* mushrooms (pp. 109–111), the poplar fieldcap is easy to grow. Slices of white wood (poplar, willow, etc.)—and even slats from produce crates made of poplar—can be dampened and rubbed with mushroom gills to form a good substrate. Then these "seeded" pieces of wood just need to be placed in a cool area (along a hedge, for example) and covered with a thin layer of dirt in order to produce—with a little luck—lovely little crops of this beloved mushroom.

- ◉ **Cap:** *up to 4 inches (10 cm) wide, fleshy and convex, smooth, white or cream-colored, sometimes brown as it ages.*
- ◉ **Gills:** *emarginate, very crowded, the same color as the cap; spore print white.*
- ◉ **Stem:** *firm, the same color as the cap.*
- ◉ **Flesh:** *firm, cream-colored; strong flavor and odor of fresh flour.*
- ◉ **Habitat:** *grass near hedges and on lawns.*
- ◉ **When to harvest:** *In the spring (rarely later).*

St. George's mushroom

Calocybe gambosa (Fr. : Fr.) Donk

The St. George's mushroom is very common in Europe and the UK but has yet to be reported from North America. It is a sought-after mushroom in many areas. It is very easy to recognize thanks to its thick flesh, its very crowded gills, and its strong scent of flour. The short length of time for which it appears also limits the risk of confusing it with other species. However, opinions differ as to how tasty it really is: some consider it to be one of the best edible mushrooms, while others find it absolutely impossible to eat. It must be said that its heavily perfumed flesh leaves little room to be indifferent.

Warning

There is another springtime mushroom that, at first glance, may be confused with the St. George's mushroom: the deadly fibrecap (p. 241) can be white or whitish, but its cap is cone-shaped and fibrillose, its gills become brown in maturity, and its flesh gives off a pleasant smell, almost similar to that of honey. In addition, it also reddens remarkably as soon as it is damaged, which is a clearly visible and distinctive feature. Certain *Entoloma* (see pp. 65, 185, 187) also grow around the same time and in the same places, but their gills have a great deal more space between them and become pink as they age.

Did you know?

Numerous studies have taken place on mushrooms around the world in order to discover new active molecules for the future of medicine, and an active substance that lowers blood sugar levels (hypoglycemic) has been discovered in the St. George's mushroom. That being said, the repetitive consumption of this mushroom is not advised—a good rule of thumb for all mushrooms—and it is therefore much too early to vaunt the therapeutic properties of the St. George's mushroom.

Mousseron or not a mousseron?

In French, the St. George's mushroom is sometimes called the "Vrai-Mousseron" (true mousseron) as opposed to the "Faux", or false mousseron, the Scotch bonnet (p. 90). In reality, neither of these mushrooms can truthfully be called a mousseron. This word was at one time used to designate all mushrooms with gills. Over time, and with the advancement of knowledge in this field, the use of this qualifier has been restricted to only a few species, in particular the two we have just mentioned.

Honey fungus

Armillaria mellea (Vahl. : Fr.) P. Kumm.

- **Cap:** *up to 6 inches (15 cm) wide, ochre-yellow and covered with fine brownish scales.*
- **Gills:** *somewhat decurrent, only slightly crowded, cream to yellow-colored, often speckled with brown; spore print white.*
- **Stem:** *fibrous, yellowish to brown, with a distinct ring that is white or yellow.*
- **Flesh:** *white, thin; sweet flavor and faint odor.*
- **Habitat:** *in clumps on living or dead deciduous trees, rare on conifers.*
- **When to harvest:** *All year long, apart from periods of severe cold; autumn in North America.*

Along with the sulphur tuft (p. 189), the honey tuft is one of the most common mushrooms that grow in clumps on wood. In mushroom season, it is hard to take a walk in the forest without running into them. Though prized in some Eastern European countries, it is not considered a good quality edible mushroom, especially because it sometimes causes small and harmless episodes of indigestion in people with sensitive stomachs. Caution is recommended when first eating this mushroom, even though it is sold in abundance in the markets of southern European countries, notably in Italy. It is often confused with the spindle shank (p. 67), an edible mushroom of equal quality that can be distinguished by its smooth cap and its spindle-shaped stem that does not have a ring.

Did you know?

Despite its inoffensive appearance, the honey fungus is a mighty parasite responsible for a disease called *Armillaria* root rot. Its underground mycelium attacks and kills tree roots and is capable of covering long distances, ravaging entire tree plantations. It may attack many kinds of trees, including those that bear fruit. It is a nightmare for arborists because few solutions exist to eradicate it.

Monk's head
mushroom

Infundibulicybe (Clitocybe) geotropa (Bull.) Harmaja

- ⊚ **Cap:** *up to 6 inches (15 cm) wide, at first with a strong umbo or nipple-shaped protrusion that quickly transforms into a funnel; the umbo remains, usually at the bottom of the funnel, ochre-beige to white.*
- ⊚ **Gills:** *decurrent, relatively widely spaced, the same color as the cap; spore print white.*
- ⊚ **Stem:** *relatively thick, fibrous, the same color as the cap.*
- ⊚ **Flesh:** *rather elastic, ochre-cream color; sweet flavor and a pleasant odor that is difficult to describe, rather refreshing, somewhat reminiscent of cut grass.*
- ⊚ **Habitat:** *under various deciduous trees, but also in grass at the edge of the forest, often in fairy rings.*
- ⊚ **When to harvest:** *Autumn until the beginning of winter.*

This large mushroom is easy to recognize, and is reported from the United Kingdom, but not from North America. It is a good quality edible, although its rather strong odor—described as "cyanic" because it is reminiscent of the acid of the same name—is not appreciated by everyone. It grows at the end of the season and endures even after the first frosts, making it, along with the wood blewit (p. 82), a mushroom worth being able to identify. It can be confused with very few other mushrooms: the clouded agaric (p. 61) grows in the same season and is also edible—though often poorly tolerated—but its convex gray cap and very crowded gills set it apart. The common funnel (*I. gibba*), reported from the UK but rarely from North America, is also edible, and is a very common miniature version of the monk's head mushroom that grows in the woods: its cap does not grow wider than 2.5 inches (6 cm) and it is less appreciated as a result of its much thinner flesh.

Warning

This *Clitocybe* often grows on roadsides, and its pale colors can easily be seen among the brown colors of late autumn. However, resist the temptation to eat it if it is growing in places that could potentially be contaminated: like all mushrooms, it tends to accumulate many pollutants and may at the very least cause indigestion. Harvest it in the forest instead.

Aniseed toadstool

Clitocybe odora (Fr.) P. Kumm

- ⊙ **Cap:** up to 3.5 inches (9 cm) wide, smooth, somewhat sunken in the middle, green, blue-green, or gray-green.
- ⊙ **Gills:** somewhat decurrent, ochre-cream or a little green; spore print white tinged pink.
- ⊙ **Stem:** short, fibrous, cream or green-cream color.
- ⊙ **Flesh:** cream-colored; very strong aniseed flavor and odor.
- ⊙ **Habitat:** under deciduous trees.
- ⊙ **When to harvest:** In autumn.

The aniseed toadstool is the only green mushroom in Europe, the United Kingdom, and North America with a strong aniseed odor, and is easy to recognize. Other *Clitocybe* release the same odor, but none of them present the same hues.

Warning

Clitocybe in the frosty funnel group (p. 177) are extremely poisonous and may grow mixed in with the aniseed toadstool: make sure to check for the presence of greenish coloring (this has a tendency to fade in older specimens) and the pure and strong smell of aniseed. See also the verdigris roundhead (p. 206).

In the kitchen

If you don't like aniseed, the aniseed toadstool is not for you. The scent of this small mushroom is so concentrated that a single piece can flavor an entire dish. Some people use it as a condiment for a plate of langoustines, while others serve it only at dessert—in ice cream, for example. Make sure not to use the stem, which is too fibrous to be appreciated.

Clouded
agaric

Clitocybe nebularis (Batsch : Fr.) P. Kumm

- ◉ **Cap:** *reaching 8 inches (20 cm) in width, convex, fleshy, gray to gray-brown or gray-yellow, often covered by a fine whitish bloom.*
- ◉ **Gills:** *somewhat decurrent, pale cream to yellow color; spore print white to pale cream.*
- ◉ **Stem:** *often club-shaped, collects leaves at its base, grayish.*
- ◉ **Flesh:** *white; sweet flavor and strong, often unpleasant odor.*
- ◉ **Habitat:** *under deciduous and coniferous trees, often in fairy rings.*
- ◉ **When to harvest:** *Autumn until the beginning of winter.*

The clouded agaric is a very common mushroom found at the end of autumn and the beginning of winter. but is not reported from North America nor the United Kingdom. It grows in colonies just about everywhere, nestled in the dead leaves and needles it feeds on. In addition to its growing season, it is also known for its "cloudy" gray cap, its somewhat decurrent gills, and its fleshy stem that is thicker at its base. Its flesh gives off a rather strong odor that is hard to describe and, for certain people, may be unpleasant.

Warning

The clouded agaric has a controversial reputation: considered a good quality edible by many mushroom lovers, we cannot forget that it has also been blamed for rather serious poisonings, though it is unclear whether these were the result of individual intolerances or proof of local variations in the mushroom's toxicity. So, if you wish to consume a clouded agaric, harvest it young and far from any sources of pollution and make sure it is properly cooked.

Sweetbread
mushroom

Clitopilus prunulus (Bolton : Fr.) Fr.

- ◉ **Cap:** *up to 4 inches (10 cm) wide, highly variable in shape, milky or grayish white.*
- ◉ **Gills:** *very decurrent, fairly crowded, whitish then distinctly pink in maturity; spore print pink to brownish-pink.*
- ◉ **Stem:** *whitish.*
- ◉ **Flesh:** *rather thick, very fragile and breaks easily; strong flavor and odor of fresh flour.*
- ◉ **Habitat:** *in deciduous and coniferous forests.*
- ◉ **When to harvest:** *In summer and autumn.*

The sweetbread is a very common and largely neglected species reported from the United Kingdom and North America; it is, nevertheless, a good mushroom to eat as long as you like its strong taste, which is unmistakable. It is also known as "the miller," probably because of its strong smell. Be certain to confirm that your collection has a pinkish spore print.

Warning

Be careful not to confuse the sweetbread with white poisonous *Clitocybe*, in particular with the frosty funnel (p. 177), which resembles it a great deal. The gills of the frosty funnel, however, are almost never decurrent and never turn pink, and its flesh—which has an odor that is more earthy than flour-like—is elastic and does not break easily. If you drop a sweetbread, only tiny pieces will be left; if you allow a frosty funnel to fall, it is unlikely to break. This seemingly simplistic test is nevertheless very informative.

Gypsy
mushroom

Cortinarius caperatus (Pers. : Fr.) Fr.

- ◉ **Cap:** *up to 5 inches (12 cm) wide, covered in the center by a fine white or lilac bloom atop an ochre-cream to brown-beige background, wrinkled as it ages.*
- ◉ **Gills:** *emarginate, cream turning to ochre-beige, denticulate edges; spore print rusty brown.*
- ◉ **Stem:** *fibrillose, whitish with a membranous white or lilac ring.*
- ◉ **Flesh:** *cream-colored; mild flavor and odor, not distinctive.*
- ◉ **Habitat:** *under deciduous and coniferous trees.*
- ◉ **When to harvest:** *Summer (in mountainous and colder regions) and autumn.*

This little-known mushroom is a wonderful edible and is reported from North America and the United Kingdom. It can be recognized by the bloom on its cap and, later, by its cap's wrinkled appearance, its particular ochre-beige coloring, its ochre gills, and finally by its stem's well-formed ring. *Agrocybe* species like the poplar fieldcap (p. 55) are edible—or at least nontoxic—and have gills that are more of a "café au lait" color than ochre. They also usually grow in grass or on wood.

Warning

Even though confusion is highly unlikely, watch out for ringed *Galerina* like the funeral bell (p. 237) that sometimes grow on the ground and are deadly.

Did you know?

In French, the gypsy mushroom's common name is the "pholiote ridée," even though it is not at all a member of genus *Pholiota*. The French name is a remembrance of the Latin name (*Pholiota caperata*) that it bore at one time before a better understanding of the mushroom placed it in its own genus (under the name *Rozites caperata*) and finally in genus *Cortinarius*. This most recent classification made the most sense when mycologists determined that the ring—which had once justified its exclusion from genus *Cortinarius*, at the time defined as mushrooms without membranous rings, but instead a cobwebby partial veil—was not a discrete feature.

Goliath
webcap

Cortinarius praestans Fr.

- ◉ **Cap:** *up to 8 inches (20 cm) wide (sometimes even larger), viscous and noticeably wrinkled along the edges as it ages, lilac-brown or crimson-brown, with white or blue veil remnants scattered here and there.*
- ◉ **Gills:** *emarginate, whitish or lilac then rusty brown; spore print rusty brown.*
- ◉ **Stem:** *thick, white or lilac, with white or blue bands from the veil.*
- ◉ **Flesh:** *white, thick; mild flavor and odor, not distinctive.*
- ◉ **Habitat:** *under deciduous trees.*
- ◉ **When to harvest:** *In autumn.*

This magnificent *Cortinarius* species astonishes with its size, its colors, and the at once wrinkled and well-dressed look of the white scales on its cap. It is an excellent mushroom to eat, but it is a rare mushroom threatened by extinction in certain areas, so mushroom lovers are therefore advised to refrain from harvesting it in order to protect its numbers. The Goliath webcap has not yet been reported from North America or the United Kingdom. Eating *Cortinarius* species in those regions is not recommended, with the exception of the gypsy mushroom.

Cortinarius balteatocumatilis, reported from the UK but rarely from North America, sometimes resembles it in terms of color and size, but it is much more common, its cap is not wrinkled, and its flesh gives off an often strong earthy odor; it is not edible.

Did you know?

Cortinarius are spread among several groups based on the appearance of the surface of their caps. Some, like the Goliath webcap and the splendid webcap (p. 235), have a viscous cap and a dry stem, others are completely dry with a felted cap, as is the case with the violet webcap (p. 184) or the deadly webcap (p. 234), for example.

Shield pinkgill
mushroom

Entoloma clypeatum (L.) P. Kumm

- ⊙ **Cap:** *up to 4 inches (10 cm) wide, convex and usually with a distinct dome-shaped umbo, brown or dark gray-brown, becomes much paler as it dries.*
- ⊙ **Gills:** *adnate to emarginate, whitish then pink; spore print rosy yellow.*
- ⊙ **Stem:** *firm and full, fibrous, grayish or brownish.*
- ⊙ **Flesh:** *whitish, white in the stem; flavor and odor distinctly flour-like.*
- ⊙ **Habitat:** *especially in hedges, under shrubs in the Rosaceae family (hawthorns, blackthorns, etc.).*
- ⊙ **When to harvest:** *Only in spring.*

Several *Entoloma* species grow in hedges in the spring, and they are not always easy to tell apart. *Entoloma saundersii* (reported from the UK), for example, strongly resembles the shield pinkgill, but it is typically fleshier, and its cap is covered in the center by a silvery gray hue that gives it a glittery sheen. *Entoloma aprile* (reported from the UK but rarely from North America), on the other hand, is distinctly sleeker, its stem is hollow, and it grows predominantly under elm trees. While nontoxic, these two species are not considered edible. The shield pinkgill mushroom is reported from the UK, but rarely from North America. It cannot be recommended as edible because it may easily be confused with toxic similar species.

Warning

The shield pinkgill, which becomes much paler as it dehydrates—it is called a "hygrophanous" mushroom for this reason—may also be confused with a dangerous poisonous mushroom that also grows in the spring: the deadly fibrecap (p. 241). Its conical cap is distinctly fibrilose, white at first then brown, its flesh reddens as it matures or when cut, and it releases an odor that is not at all flour-like but instead fruity and similar to the scent of honey.

Golden needle mushroom or
Velvet foot

Flammulina velutipes (L.) P. Kumm

- ⊙ **Cap:** *up to 2.5 inches (6 cm) wide, viscous and sticky, yellow-orange to red-orange.*
- ⊙ **Gills:** *adnate, not very crowded, white then ochre-beige; spore print white.*
- ⊙ **Stem:** *fairly tough, black velvety all over, yellow at the top, and dark brown going toward the base.*
- ⊙ **Flesh:** *cream-colored; mild flavor and odor.*
- ⊙ **Habitat:** *in clumps on dead deciduous wood, typically elm.*
- ⊙ **When to harvest:** *At the end of autumn and even in winter into spring.*

*E*nokitake . . . that's the name the Japanese give the golden needle. They grow it in large quantities for cooking, but the cultivated specimens bear little resemblance to the mushroom's "wild" forms. When it is grown in containers in the dark, the mushroom develops oversized stems with tiny caps and remains completely white; the clumps of cultivated mushrooms end up looking more like plates of spaghetti! The golden needle is an interesting mushroom because it grows during a period when many edibles have yielded to the first chill of winter. It is not often reported from North America or the United Kingdom, perhaps because it fruits in the cold season, when few are out looking.

Warning

There are other *Flammulina* that exist, and fortunately they are all edible, but they are difficult to distinguish from the golden needle without the use of a microscope. *Flammulina elastica* (rarely reported from either North America or the UK) and *F. populicola* (reported from North America but not the UK), for instance, are only differentiated by the subtle differences in their spores and environmental preferences for certain kinds of wood (willows and aspens for the former, only poplars for the latter).

Spindle shank
mushroom

Gymnopus fusipes. (Bull. : Fr.) S.F. Gray

- ◉ **Cap:** *up to 3 inches (8 cm) wide; smooth, reddish brown to a pale ochre-brown, often speckled with rusty brown.*
- ◉ **Gills:** *emarginate, thick and far apart, ochre-cream color, often speckled with rusty brown; spore print white.*
- ◉ **Stem:** *typically spindle-shaped and twisted, rooting, cream at the top and brown-red going toward the base.*
- ◉ **Flesh:** *cream-colored; flavor and odor of mushroom, not distinctive.*
- ◉ **Habitat:** *in clumps on deciduous tree logs.*
- ◉ **When to harvest:** *Summer and autumn.*

These mushrooms, reported from the UK but rarely from North America, grow in clumps on logs and sometimes at the base of tree trunks, which is already a great help in identifying them. The spindle shank's stem is also very distinctive: it is quite twisted and terminates in a long point that penetrates the dead wood—unless that point is broken while it is being harvested—and is almost two-tone, pale at the top and reddish-brown at the bottom. Even though it is sold widely in the markets of southern Europe, it is not always very well-liked. It has been blamed for mild and harmless cases of poisoning that were more likely caused by fungi or bacteria growing on the spindle shank than the spindle shank itself. Because of this it has been recently reported as inedible or toxic; caution is advised.

Did you know?

While this species is often found on logs, it can also develop on the roots of weakened living trees and cause a disease called *Collybia fusipes* root rot. Less virulent than its genus *Armillaria* counterparts, it is nevertheless capable of hastening a tree's decay and will continue to feed on its dead wood once that tree has died.

- ◉ **Cap:** up to 2.5 inches (6 cm) wide, and 2 inches (5 cm) to 8 inches (20 cm) in height, initially in the shape of a "gloved finger" then transforming into a bell, covered with white scales that become gradually darker approaching the edge.
- ◉ **Gills:** very crowded, white then pink and finally black, liquifying at the edge (transforming into "ink"); spore print black and wet.
- ◉ **Stem:** long, white, with a white ring all the way at the bottom.
- ◉ **Flesh:** white, but quickly turns to liquid; faint and pleasant flavor and odor.
- ◉ **Habitat:** grass in fields, lawns, roadsides, etc.
- ◉ **When to harvest:** All year long apart from periods of severe cold, but especially in autumn.

Shaggy inkcap, Shaggy mane, or Lawyer's wig

Coprinus comatus Fr. : Fr.

The shaggy inkcap is very common in North America and the UK, is particularly easy to recognize, and is widely harvested. With its cap that is taller than it is wide and covered in white scales, and its gills that turn black and tend to transform into an unappetizing black liquid, it is hard to mistake for anything else. In mature specimens, the gills dissolve and the entire cap usually disappears. It is for this reason that we sometimes see clumps of "posts" smeared with black ink, vestiges of stems that remain long after the caps they used to carry.

Deliquescence

Coprinus species, for the most part, are all capable of rapidly turning to liquid. This property is called deliquescence, and these mushrooms are therefore deliquescent. The "ink" that results from this liquefaction is black because it is colored by millions of miniscule spores formed on the gills. Some mycologists have used this ink for writing, and it seems to hold up over time. The shaggy inkcap is also sometimes called "goutte d'encre" or "drop of ink" in French, referring to a use for this mushroom that is more anecdotal than traditional.

Warning

There are other much rarer *Coprinus* species that resemble the shaggy inkcap. Whether or not they are edible is unknown. *Coprinus levisticolens* (not reported from North America or the UK) is an almost perfect lookalike that grows on sandy ground, but it gives off a strong chicory scent. Avoid confusing the shaggy inkcap with the magpie inkcap (p. 183), which has a black cap covered with white scales, grows in the woods, and gives off an unpleasant odor of asphalt.

Did you know?

Until a few years ago, genus *Coprinus* consisted of a large group of around 150 species in Europe. Careful scientific studies attempting to retrace species evolution and to construct a classification system that takes this history into account have shown that the genus is actually made up of at least four distinct groups, one affiliated with the *Agaricaceae* (genus *Coprinus*, which includes the shaggy inkcap), and the rest related to the *Psathyrellaceae* (see p. 200). For mycologists, the consequences of this discovery have been revolutionary, not least because ninety-five percent of *Coprinus* species now have new Latin names as a result.

▲ *Gill maturation in the shaggy inkcap.*

Slimy
spike cap

Gomphidius glutinosus (Schaeff. : Fr.) Fr.

- ⦿ **Cap:** *up to 4 inches (10 cm) wide, covered by a viscous veil that peels easily, chocolate brown to lilac-gray, sometimes stained black in places.*
- ⦿ **Gills:** *decurrent, spaced, white then gray and finally black; spore print smoky gray to black.*
- ⦿ **Stem:** *white at the top and yellow at the base with a small blackish ring area.*
- ⦿ **Flesh:** *whitish; mild flavor and odor.*
- ⦿ **Habitat:** *under spruce and red pine trees, particularly in mountainous or cold regions.*
- ⦿ **When to harvest:** *Autumn.*

This viscous mushroom (reported from the UK and North America) with a bright yellow base that grows under spruce or red pine trees is not difficult to identify. It is consumed in certain regions, but its soft and bland flesh is not for everyone and the viscous layer on the cap should be peeled. If nothing else, it grows in colonies, which makes it an easy mushroom for harvesting. The larch spike cap, *Gomphidius macu-* *latus* (reported from North America and the UK), resembles it somewhat, but the base of its stem is not yellow, it turns black on all of its parts, and it only grows under tamarack trees.

Did you know?

Spore color, observed in a mass deposit called a spore print, is a critically important criterion in mushroom identification. Though it is often necessary to conduct a spore print to know exactly what color the spores are (see p. 25), it is sometimes possible to get a very precise idea simply by attentively observing the mushrooms themselves: in slimy spike caps, for example, some of the spores that fall from the gills are trapped by the viscous bulge at the top of the stem, causing it to be colored black, the same color as the spore print.

March
mushroom

Hygrophorus marzuolus (Fr. : Fr.) Bres.

- ⊙ **Cap:** up to 6 inches (15 cm) wide, somewhat viscous, completely white when it first comes out of the ground, then gray-black to black.
- ⊙ **Gills:** decurrent, spaced, whitish then gray and often black in old specimens; spore print white.
- ⊙ **Stem:** firm, white then grayish.
- ⊙ **Flesh:** firm, white then gray; sweet flavor and fruity odor.
- ⊙ **Habitat:** under conifers, but also under beech trees; most often in mountainous areas.
- ⊙ **When to harvest:** Late winter into spring.

At the very beginning of spring, mushroom enthusiasts start keeping watch for the first signs of growth. While this time of year is a little early for morels, the March mushroom, reported only from western North America and not from the UK, is already beginning to pierce the ground in coniferous forests and occasionally beneath beech trees. Its cap is typically very pale early on, then changes quickly to gray and later a blackish color. Its decurrent gills (spaced far apart and often gray) as well as its firm stem and flesh round out its characteristic features. In certain regions it may grow beginning in February and un-

til May. While it is rather common in mountainous regions, it can also be found in lowland areas.

Warning

There are several *Hygrophorus* species growing in autumn that resemble the March mushroom, so be sure to only harvest it in the spring. The almond woodwax (*H. agathosmus*) (reported from North America and the UK, and regarded as a poor edible), for example, is commonly seen beneath coniferous trees, particularly pines, until late in the season. It can be identified by its cap, which

is more gray than black and sticky, and most of all by its pleasant bitter almond smell, also known as the "white glue" scent!

Meadow
waxcap

Hygrophorus (Cuphophyllus) pratensis (Pers. : Fr.) Murrill.

- ◉ **Cap:** *up to 2.7 inches (7 cm) wide, convex and fleshy, often with a large dome-shaped umbo, orange to ochre-yellow.*
- ◉ **Gills:** *decurrent, thick and spaced, whitish to ochre-cream; spore print white.*
- ◉ **Stem:** *fairly short, the same color as the gills or just about.*
- ◉ **Flesh:** *thick, pale orange in the cap, white in the stem; mild odor and flavor.*
- ◉ **Habitat:** *usually grass in unfertilized meadows and fields.*
- ◉ **When to harvest:** *Autumn until the beginning of winter.*

This easy to recognize mushroom, reported from North America and the UK, is sometimes confused with the Scotch bonnet (p. 90), which grows in the same areas but is far less fleshy and has emarginate gills. Any ensuing error in identification is harmless because the Scotch bonnet is a perfectly good mushroom to eat. An almost perfect meadow waxcap lookalike, *H. nemoreus* (reported from the UK, but very rarely from North America), is sometimes seen in forests. It is also edible and its gills are more crowded, however, its cap more fibrillose, and most significantly, its flesh gives off a distinct odor of flour.

Did you know?

Many *Hygrophorus* species are difficult to observe. In Europe, they only grow in places that are completely devoid of pollution, and particularly in old and untouched natural prairies. With the extension of agricultural zones, areas like this are becoming smaller and smaller in size, and are taking their *Hygrophorus* species with them, making these mushrooms increasingly rare. It is therefore important to protect these species and to preserve the places where they grow, for these areas are home to many other organisms also on the path to increasing scarcity.

Sheathed woodtuft
mushroom

Kuehneromyces mutàbilis (Scop. : Fr.) Sing & A.H. Sm.

- ◉ **Cap:** *up to 2.4 inches (6 cm) wide, russet brown or cinnamon brown at first, then becomes much paler as it dries out; the dehydration begins at the center of the cap and continues in a rosette pattern.*
- ◉ **Gills:** *adnate or somewhat emarginate, cream-colored then brown; spore print rusty to cinnamon brown.*
- ◉ **Stem:** *reddish-cream color, covered with fine, upturned scales beneath a small and labile membranous ring.*
- ◉ **Flesh:** *brownish; sweet flavor and faint odor.*
- ◉ **Habitat:** *in clumps on deciduous and coniferous tree stumps.*
- ◉ **When to harvest:** *Autumn.*

The sheathed woodtuft is well-known in certain regions, and is often harvested under the name of "Agaric à soupe," or soup agaric. Its cap dehydrates in a rather distinctive manner, but the most attention should be paid to its stem: be sure to note the small scales that cover its lower two-thirds. This armilla, or superior remnant of the universal veil, as mycologists call this kind of covering, is a distinguishing feature: if you are not sure you can see it or if the specimens are too old to be closely scrutinized, don't take the risk, because the possibility for confusion with other species is significant.

Warning

The funeral bell (p. 237) is deadly and strongly resembles the sheathed woodtuft. It grows on decaying wood, but more in colonies than in clumps: the stems of the mushrooms are not fused. Its cap is similarly colored and grows pale just as rapidly. Its stem, however, bears a small ring and is fibrillose, without a single scale. The skin of the cap, which is slightly larger than the cap itself, hangs over creating a thin transparent border around its periphery. The sheathed woodtuft is reported from North America and the UK, but because of its similarity with the deadly funeral bell, it cannot be recommended for consumption, except for the very observant and experienced.

Amethyst
deceiver

Laccaria amethystina (Huds.) Cooke

- ◉ **Cap:** *up to 2.5 inches (6 cm) wide, matte and velvety, purple to bright blue-purple, dehydrates into whitish lilac.*
- ◉ **Gills:** *adnate, spaced, purplish; spore print white.*
- ◉ **Stem:** *fibrous, purplish or lilac-brown, fades as it dries.*
- ◉ **Flesh:** *thin; mild flavor and odor.*
- ◉ **Habitat:** *under deciduous and coniferous trees.*
- ◉ **When to harvest:** *Summer through autumn.*

This mushroom reported from across North America and in the UK is largely neglected by many enthusiasts but nevertheless deserves to be better known. It often grows in abundance and its flesh is far from bland. Admittedly easier to identify when it is well-hydrated, it becomes unrecognizable when dried out, which increases the risk of confusing it with something else. Nevertheless, its lilac hues, spaced gills, and odorless flesh are excellent clues that may prevent you from making a mistake. There are several other *Laccaria* species that are often difficult to tell apart, but none of them are as vibrantly colored as the amethyst deceiver.

Warning

The amethyst deceiver is primarily confused with the lilac bonnet (p. 197), which is extremely poisonous and sometimes sports the same colors. The lilac bonnet's cap, however, is smooth, rather glossy, and its gills are much more crowded. Its flesh also gives off a strong radish smell that allows it to be recognized beneath its many disguises.

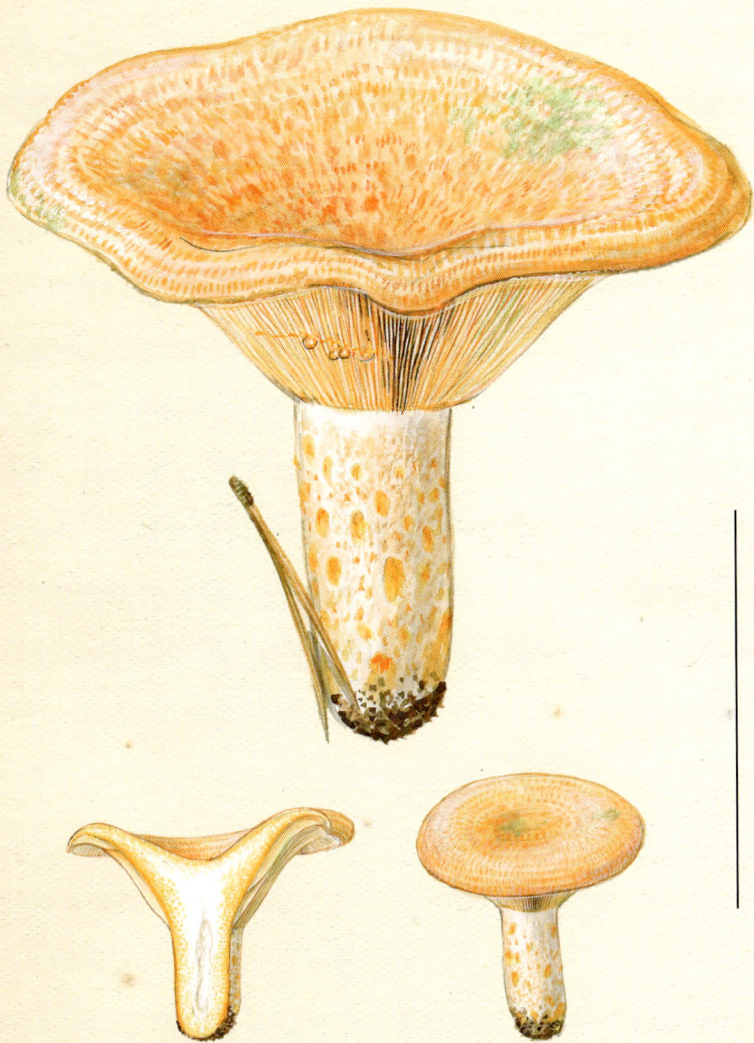

Saffron milkcap

Lactàrius deliciosus (L. : Fr.) S. F. Gray

- ◉ **Cap:** *up to 6 inches (15 cm) wide, sunken in the center, orangey-salmon to orangey-brown with darker concentric circles, turns green as it matures.*
- ◉ **Gills:** *somewhat decurrent, orangey-salmon then stained with green; spore print ochre-cream.*
- ◉ **Stem:** *hollow and brittle with darker dimples.*
- ◉ **Flesh:** *brittle, cream or orange-colored; exudes a moderate amount of bright orange milk; mild flavor and odor, a little fruity.*
- ◉ **Habitat:** *only under conifer trees.*
- ◉ **When to harvest:** *Autumn.*

The saffron milkcap—often improperly called the Catalan, a name technically reserved for the bleeding milkcap (p. 79) reported from the UK and western North America—is a good edible with firm flesh as long as it is harvested young before it has turned green. As it ages it tends to quickly fill with worms. It is easy to recognize if we keep in mind its habitat. There are other *Lactarius* species that have milk and orange flesh, but they usually grow associated with other conifers. This is not the case for the carrot milkcap (*L. quieticolor*) (reported from the UK), however: though a lesser-quality edible, it grows happily under pine trees. Its orangey to grayish-brown cap is duller than that of the saffron milkcap and its stem has only a few dimples or none at all. A similar species, *L. deterrimus*, also edible and found beneath conifers, is reported from the UK and North America.

In the kitchen

The saffron milkcap lends itself to a variety of preparations. If harvested young, it is sometimes preserved in vinegar with a little bit of honey and aromatics and then eaten like a pickle. When harvested a little older, it can easily be used in a sauce or omelette provided one appreciates mushrooms with firm flesh, though this is not necessarily something everyone enjoys.

Velvet
milkcap

Lactarius lignyotus Fr.

- ◉ **Cap:** *up to 4 inches (10 cm) wide, often with an umbo, wrinkled, dark chestnut brown to dark brown.*
- ◉ **Gills:** *decurrent, white then cream, contrasting sharply with the cap and the stem.*
- ◉ **Stem:** *hollow and brittle, brown like the cap; spore print ochre.*
- ◉ **Flesh:** *brittle, white, white milk, small amount, unchanging when cut, staining the flesh pinkish-orange.*
- ◉ **Habitat:** *under spruce and fir trees in damp or mountainous areas.*
- ◉ **When to harvest:** *Mid-summer through autumn.*

It is always a pleasure to encounter the velvet milkcap (reported from North America but rarely from the UK) under spruce trees in mid-mountain and high mountain areas or colder regions. Its cap and dark brown stem, both wrinkled, stand in stark contrast to its pale gills, forming a very elegant-looking mushroom. It may be confused with *Lactarius picinus*—reported from the UK and inedible without being poisonous— which grows in the same environment, but the latter's cap is a lighter brown, not wrinkled, lacks an umbo, and has cream-colored gills and a milk that slowly turns pink when exposed to open air. It must be noted that these two *Lactarius* species are, it seems, on the path to becoming quite rare in Europe; protecting them is therefore recommended. There are several similar species in North America; mycologists have yet to define them, but it seems that the true European species does not occur in North America.

Did you know?

Mushrooms in genus *Lactarius* and genus *Russula* (p. 95) are very closely related and are the only mushrooms with caps and stems whose flesh is made up almost entirely of round cells called sphaerocysts arranged side by side. These are the cells responsible for the texture of these mushrooms' flesh, which is firm but brittle. As their name indicates, *Lactarius* species release a milk when broken open, while *Russula* species never exhibit this feature.

Fishy milkcap/
Weeping milkcap

Lactarius volemus (Fr. : Fr.) Fr.

- ◎ **Cap:** *up to 6 inches (15 cm) wide, matte, orangey-brown to bright reddish-brown, tends to crack as it matures.*
- ◎ **Gills:** *cream to ochre color, stains reddish-brown in bruised areas; spore print whitish.*
- ◎ **Stem:** *hollow and brittle, cream to orangey-cream color, stains like the gills.*
- ◎ **Flesh:** *brittle, white, releases significant amount of white milk that turns brown when dry; sweet flavor and odor of cooking shellfish or artichoke.*
- ◎ **Habitat:** *under deciduous and coniferous trees.*
- ◎ **When to harvest:** *Summer and autumn.*

Few *Lactarius* species exude as abundant an amount of milk as the fishy milkcap (reported from eastern North America and the UK). One need only scratch the gills with a fingernail and milk will start pouring out in fat white drops. Its flesh also gives off a strong and rather pleasant odor, similar to the scent of artichokes or, more precisely, that of Jerusalem artichokes, though this reference will probably not mean very much to the young among us. There is a fishy milkcap lookalike called *Lactarius rugatus* (rarely reported from the UK, but not from North America) that grows in hot regions; not only is it slightly smaller, but its cap wrinkles rapidly as it grows, and its flesh is distinctly less fragrant. *Lactarius hygrophoroides*, edible and reported from eastern North America, also looks like the fishy milkcap but is much smaller and has very widely spaced gills. *Lactarius corrugis*, edible and also reported from eastern North America, has a wrinkled cap.

Did you know?

In order to identify mushrooms, mycologists use somewhat toxic substances that produce colorful chemical reactions when applied to the correct area. Iron salts, for example, which appear in the form of greenish crystals, allow us to easily distinguish between the fishy milkcap and *Lactarius rugatus*: when rubbed on the stem, a crystal of this reactive substance will turn the surface of the fishy milkcap olive green and the surface of *Lactarius rugatus* an orangey-pink.

- ⊙ **Cap:** up to 4 inches (10 cm) wide, hollowed in the center, pinkish orangey-brown, with or without concentric circles, turns green with age.
- ⊙ **Gills:** somewhat decurrent, pale winy pink to winy red color.
- ⊙ **Stem:** hollow and brittle, pinkish-gray, with reddish-orange to winy brown dimples.
- ⊙ **Flesh:** brittle, winy brown, exudes a small amount of milk, blood red to winy red; faint flavor and odor.
- ⊙ **Habitat:** only under pine and conifer trees.
- ⊙ **When to harvest:** Autumn.

Bleeding
milkcap

Lactarius sanguifluus (Paulet) Fr.

The bleeding milkcap is quite common in most European countries but is rarely reported from North America or the UK. It is easy to recognize because of its beautiful red milk. It sometimes appears with a different coloration in its *vinosus* form, which can be identified by its muted orange hues, purplish-pink gills, and a stem covered in a fine whitish bloom. Reported from North America and the UK, *L. sanguifluus* is very close to the *vinosus* form. It may be confused with the saffron milkcap (p. 75), which also grows under pine trees, but the saffron's milk is orange; in any case, confusing these two mushrooms poses no danger.

In the kitchen

The bleeding milkcap can be cooked in a variety of ways and you should feel free to give your imagination free rein. It should, however, be harvested young and fresh because it quickly becomes soft and wormy. It is traditionally fried in olive oil with a little bit of garlic and parsley, but like the saffron milkcap it is also pickled in vinegar to make an excellent snack. Canning is the best way to store them for several months.

Did you know?

The process of evolution has no shortage of imagination when it comes to baffling naturalists, and it has sometimes given rise to *Lactarius* species that are quite strange: in South America, for instance, certain *Lactarius* look like *Pleurotus* species and grow without stems, grafted onto heavily decayed tree trunks. Others are subterranean and look an awful lot like truffles. Still others have been imaginative enough to exchange their gills for tubes similar to those in boletes. The microscopic features specific to the *Lactarius*, however, are well-preserved, and the flesh of each of these unique varieties still exudes a milk when it is broken, regardless of the mushroom's shape.

Faithful partners...

All Lactarius *species are associated with trees via their underground mycelium: they are mycorrhizal mushrooms. This intimate relationship, put in place over the course of evolution, has achieved a high degree of specificity. In the* Lactarius *group with orange or red milk, for instance, the bleeding milkcap only associates with pines, the* Lactarius salmonicolor *(p. 192) only with fir trees, the false saffron milkcap (p. 190) only with spruces, etc. It is therefore very important when harvesting mushrooms to pay particular attention to the specimen's environment and to develop a basic level of botanical knowledge in order to identify the different trees that grow in your region.*

Shiitake

Lentinula edodes (Berk.) Pegler

- ◉ **Cap:** *up to 4 inches (10 cm) wide, convex, fleshy, covered in scales and whitish or beige fibrils on a brown-gray or dark brown background with an outer edge that remains rolled under for a long time.*
- ◉ **Gills:** *emarginate, white to grayish-cream; spore print white.*
- ◉ **Stem:** *fairly leathery, brownish, covered by fibrils and scales that are white or brown; sometimes has a small ring area.*
- ◉ **Flesh:** *white; sweet flavor, a little peppery, and pleasant odor.*
- ◉ **Habitat:** *on various deciduous trees.*
- ◉ **Cultivated extensively:** *around the world.*

Cultivated for at least 1,000 years in China and Japan, the shiitake is widely used in Asian cuisine, particularly for vegetarian dishes. It is sold fresh or dried in European markets, sometimes accompanied by other mushrooms in what are called forest blends, in spite of the fact that it cannot be found anywhere near European forests. Wild *Lentinula* do not exist in Europe, North America, or the UK, though there are reports that some may be colonizing in the areas where shiitake are cultivated in forest environments.

Warning

If you buy fresh shiitake, be sure to cook them properly and do not consume them raw under any circumstances: they contain a toxin called lentinan that is destroyed by heat but can cause an allergic reaction in some people one to three days after consuming that causes an itchy rash resembling whiplash marks. Flagellate dermatitis is well-known in Asia and has already been observed several times in Europe, North America, and the UK.

For the same reason, be careful of condiments that use shiitake as a base ingredient because these can sometimes be rich in lentinan.

Lepista luscina

Lepista luscina (Fr. : Fr.) Singer

- ◉ **Cap:** *up to 4 inches (10 cm) wide, convex, fleshy, gray-beige or gray-brown, often with darker dimples.*
- ◉ **Gills:** *adnate to emarginate, grayish-beige; spore print pinkish.*
- ◉ **Stem:** *rather fibrous, grayish-beige.*
- ◉ **Flesh:** *fairly thick, whitish; sweet flavor and a mushroom odor that is not very strong.*
- ◉ **Habitat:** *in groups in prairie and meadow grasses, especially on mountains.*
- ◉ **When to harvest:** *Summer and autumn.*

*L*epista luscina, reported from the UK but rarely reported from North America, is a species that mushroom enthusiasts in the mountains know well. It is rarely found in lowland areas but can sometimes be found at medium altitudes in prairies with low levels of pollution. Its gray cap, often marked with darker dimples sprinkled concentrically on top of it, its grayish beige gills, and its environment make it a mushroom that is relatively easy to identify. It may be confused with certain *Melanoleuca* that grow in the same habitat, but the latter, which are nontoxic, have gills that are far more crowded and usually whiter.

Warning

Lepista luscina may also be confused with the livid pinkgill (p. 187) and with some nonscaly forms of the tiger tricholoma (p. 209). These two very toxic mushrooms grow in forests and have flesh that releases a floury odor; the former has gills that go from the color of butter to pink, while the latter has gills that are usually white with a green-blue tinge.

Wood blewit

Lepista nuda (Bull. : Fr.) Cooke

- ◉ **Cap:** *up to 6 inches (15 cm) wide, convex, fleshy, lilac-blue to brownish-blue.*
- ◉ **Gills:** *adnate to emarginate, lilac-blue, easily separated from the flesh of the cap; spore print pinkish-buff.*
- ◉ **Stem:** *fleshy, the same color as the cap, with a bulbous base, fibrillose, and has a bloom.*
- ◉ **Flesh:** *bluish; sweet flavor and pleasant odor that is difficult to describe.*
- ◉ **Habitat:** *grass in forests and parks, sometimes along the edges of trails.*
- ◉ **When to harvest:** *Autumn until the beginning of winter, even after the first frosts.*

Here is a mushroom reported from across North America and the UK that is easy to identify and also has the advantage of growing at the end of the season, not unlike the golden needle (p. 66). The wood blewit can even resist the first frosts, and it is not rare to en-counter partially frozen specimens. Its rather strong flavor is not liked by everyone. It can easily be confused with the sordid blewit (p. 84), which is smaller and less fleshy, but just as edible. The field blewit (see next page) on the other hand is easily distinguished by its cap completely free of blue coloring and its distinctly purple stem. Make a spore print to ensure that it is not one of the purple *Cortinarius*, which will have a rusty-brown spore print.

Did you know?

Cultivation of the wood blewit is en-tirely feasible today. In spite of this fact, it is only rarely sold in markets because it is not well-known and, as a result, does not sell well when fresh. Nevertheless, it is beginning to find its place in canned goods and dried mushroom products used for cooking.

Field blewit

Lepista saeva (Fr.) P. D. Orton (a synonym for *L. personata* [Fr.] Cooke)

- ⊙ **Cap:** *up to 5 inches (12 cm) wide, convex, fleshy, pale brown to gray-brown, becoming paler as it dries.*
- ⊙ **Gills:** *adnate to emarginate, beige-cream, easily separable from the flesh of the cap; spore print pinkish-buff.*
- ⊙ **Stem:** *fleshy and often short, usually at least partially bright purple, rarely grayish-brown.*
- ⊙ **Flesh:** *whitish; sweet flavor and fruity odor that sometimes becomes unpleasant in maturity.*
- ⊙ **Habitat:** *grass in fields and parks.*
- ⊙ **When to harvest:** *Autumn.*

The field blewit is not common and prefers calcareous soil but is reported from North America as well as the UK. When a location is suitable, it can grow in large numbers. It can easily be identified by its very fleshy and rather pale cap, along with its beautiful purple stem that contrasts with its gills. The stem's purple shades sometimes disappear, and it then has a less typical appearance, but there is little risk of confusing it with a poisonous lookalike. As with the wood blewit, be careful that you don't have a *Cortinarius* which will have a rusty-brown spore print.

Did you know?

The gills in many *Lepista* species easily detach from the flesh of the cap—sometimes all it takes is being lifted up by a fingernail—and their spores are generally pink. This detachable gill feature is also found among *Paxillus* species, which have a brown spore print, but because of this similarity, *Lepista* were once classified under genus *Rhodopaxillus*, from *rhodos*, "pink," a reference to the color of the spores.

Sordid blewit

Lepista sordida (Schumach. : Fr.) Singer

- ⊙ **Cap:** up to 4 inches (10 cm) wide, flat or nearly flat, not very fleshy and often striated at the edges, lilac or purplish to gray-brown lilac, becoming paler as it dries.
- ⊙ **Gills:** adnate to emarginate, whitish to lilac, easily separated from the flesh of the cap; spore print pink to salmon.
- ⊙ **Stem:** not very fleshy, just about the same color as the cap.
- ⊙ **Flesh:** bluish, thin; sweet flavor and faint, pleasant odor.
- ⊙ **Habitat:** in forests and in parks, especially under deciduous trees.
- ⊙ **When to harvest:** Autumn until the beginning of winter, even after the first frosts.

This is the wood blewit's younger brother (p. 82) (reported from the UK, but rarely from North America), but the two are not always easy to tell apart. The mycologist's "trick" for differentiating the two species is to turn the cap toward a source of light as if you were trying to look through it: if it is totally opaque, you have probably stumbled upon a wood blewit, but if it is a little bit translucent it is most likely a sordid blewit. These two *Lepista* are both edible, so confusing them does not put you in any danger.

Did you know?

The sordid blewit's colors are remarkably variable: it can be a bright and deep purple or, on the contrary, totally devoid of a single blue pigment. In certain cases, this makes it hard to identify, and mycologists have described many forms and varieties to do their best to pin down this multifaceted mushroom.

White
dapperling

Leucoagaricus leucothites (Schumach. : Fr.) Singer

- ⊙ **Cap:** *up to 6 inches (15 cm) wide, convex, often with a large dome in the center, dry and matte, white or cream-colored.*
- ⊙ **Gills:** *free, white then cream, sometimes pink as it ages; spore print white.*
- ⊙ **Stem:** *often club-shaped, white, with a small ring that slides a little bit.*
- ⊙ **Flesh:** *white; sweet flavor and faint odor.*
- ⊙ **Habitat:** *on lawns and in prairie grasses.*
- ⊙ **Where to harvest:** *Late summer through autumn.*

Mushroom enthusiasts searching for the meadow mushroom (p. 45) are often bewildered by the white dapperling (reported from North America and the UK) because it looks a lot like their favorite mushroom from above, but it has white gills instead of pink or brown ones. This is why the white dapperling is often found thrown in the grass by some disappointed harvester. Though it may be less renowned than the meadow mushroom, it is nevertheless an entirely honorable edible.

Warning

There is a significant risk of confusing the white dapperling with deadly white *Amanita* (see pp. 40–41): they have the same white cap, the same free gills, also white, and the same ring on the stem. Never harvest a white dapperling too close to trees that an *Amanita* might associate with—at the edge of the woods, for example—and make sure that the ring can slide and that, most importantly, there is no volva hiding at the base of the stem. Confusion with the yellow stainer (p. 167) is less serious and less likely because it has pink gills that turn brown and very yellow flesh. There have been reports of gastric problems from the gray form of this mushroom.

- **Cap:** *up to 8 to 10 inches (20–25 cm) wide, beginning as a ball on top of the stem—the whole ensemble resembling a drumstick—that flattens out, covered in fairly large brown scales on a cream background.*
- **Gills:** *free, white or cream, forming a smooth circle around the top of the stem; spore print white to pale pink.*
- **Stem:** *bulbous, covered by brown flecks on top of a cream background with a thick sliding ring.*
- **Flesh:** *fairly fibrous, white; sweet flavor and strong odor that is either herbaceous or similar to melted butter.*
- **Habitat:** *in fields and prairies or in sparsely wooded areas.*
- **When to harvest:** *Summer and autumn.*

Parasol
mushroom

Macrolepiota procera (Scop. : Fr.) Singer

Leave old parasols behind!

Like other large edible Lepiota species, the parasol mushroom is relatively fibrous, particularly when it is aging. Some people with digestive issues cannot tolerate its long fibers, which can sometimes accumulate in the intestines and cause a blockage called a mushroom bezoar. To avoid this nuisance, only harvest young specimens that are not fully unfurled and whose gills are still white. If you have recurrent digestive issues, avoid this species entirely or only eat it when it is very young and well-cooked.

While some mushrooms are hard to make out in their natural habitat, this is definitely not the case for the parasol; it can easily reach 12 inches (30 cm) in height and the most developed specimens could easily serve as children's umbrellas. There are few gilled mushrooms in the United Kingdom that can reach such proportions. With its large size, its cap covered in brown scales against a pale background, and its stem with a ring that slides just like a piece of

jewelry, it is certainly easy to identify. Still, remember to watch out for the small and deadly *Lepiota*.

This species is reported from across the United Kingdom. There are similar species reported from North America that have been called *procera*, that are equally as flavorful, but they are different enough that eventually there will be new species names for them.

In the kitchen

The parasol should be harvested when it is still young and when its cap has not yet fully blossomed (see previous page). The stem must be left out of cooking preparations entirely because it is simply too tough and fibrous. The caps can be prepared in multiple ways: fried, in a cream or tomato sauce, breaded, or, according to one well-known French recipe, placed flat on a grill, gill side up,

and seasoned with olive oil, pepper, and a persillade (parsley) sauce.

Warning

Certain *Lepiota* are known for being very poisonous, even deadly, and while the risk of confusion with the parasol is low (except in warmer parts of North America where the equally large green gilled *Lepiota* causes gastric problems every year), this still presents problems for mushroom harvesters. How can you avoid making a mistake? First of all, none of the small poisonous *Lepiota* will reach the size of the parasol: the largest of them have caps that will reach 4 inches (10 cm) wide at most. Furthermore, none of them possess a sliding ring like the parasol's: in their case, the ring is connected to the surface of the stem and any attempt to move it would cause it to break. Second, always verify that none in your collection have a greenish spore print. Do not make the mistake of thinking an *Amanita* is a *Macrolepiota*.

Shaggy parasol
mushroom

Chlorophyllum rhacodes (Vittad.) Singer

- ◉ **Cap:** *up to 6 inches (15 cm) wide, covered with drab grayish-brown and upturned scales on a background that is roughly the same color.*
- ◉ **Gills:** *free, white or cream, redden when bruised; spore print white.*
- ◉ **Stem:** *bulbous, smooth, reddens significantly when bruised with a thick sliding ring.*
- ◉ **Flesh:** *white, reddens when bruised; sweet flavor and faint odor.*
- ◉ **Habitat:** *in and around sparsely wooded areas under coniferous and deciduous trees.*
- ◉ **When to harvest:** *Autumn.*

With its "badly combed" curly scales and dark hues, the shaggy parasol, reported from North America and the United Kingdom, is aptly named. One need only scratch the surface of the stem with a fingernail to see the flesh turn its typical beautiful madeira red in just a few minutes. It is a good edible that should only be eaten when it is young because, like the parasol (p. 87), its fibrous flesh may be poorly tolerated by certain people. Like many *Lepiota*, it favors places that are somewhat polluted and rich in nitrogenous compounds; only harvest it when it is growing in places free from any major sources of contamination.

Warning

The greatest risk for confusion comes from the *Chlorophyllum brunneum* (p. 176), reported from the UK and North America, which shares many of the shaggy parasol's characteristics: it too is a large *Lepiota* with a sliding ring whose flesh reddens where it is bruised. To distinguish this large inedible *Lepiota* from the shaggy parasol, remember that its cap's brown scales rest against a pale, whitish background in greater contrast than the brownish scales on the shaggy parasol, which blend in almost completely with the base color of the cap.

Weeping
slimecap

Limacella guttata (Pers. : Fr.) Konrad & Maublanc

- ⊙ **Cap:** *up to 5 inches (12 cm) wide, viscous, pinkish ochre-beige.*
- ⊙ **Gills:** *free, cream-colored; spore print white.*
- ⊙ **Stem:** *often club-shaped, cream, with a large skirt-shaped ring that exudes olive-gray droplets in humid weather.*
- ⊙ **Flesh:** *white; sweet flavor, floury in texture, strong odor of flour or cucumber.*
- ⊙ **Habitat:** *in deciduous and coniferous forests, most often in the mountains.*
- ⊙ **When to harvest:** *Autumn.*

The weeping slimecap, reported from the UK but rarely from North America, is easily identified once you have taken a careful inventory of its characteristic features: its generalized cream coloring, its viscous cap, its ring that "weeps" olive-colored droplets, and its strong cucumber odor. The dripping slimecap (*L. illinita*) (reported from North America, but not the UK), edibility unknown, resembles it somewhat, but its stem does not have a ring. The weeping slimecap also has a silhouette similar to that of the gypsy mushroom (p. 63), but the gypsy is not viscous, its cap has a bloom and tends to wrinkle, and its gills become ochre-brown in maturity.

Warning

As is the case with all gilled mushrooms with a ring, the risk of confusing them with deadly *Amanita* cannot be overlooked. It should be noted that no deadly *Amanita* gives off a strong flour or cucumber odor like the weeping slime cap, and all deadly *Amanita* emerge from a volva that remains at the base of the stem in the form a membranous sac. Of course, you will need to consider the other characteristics to determine the edibility of your collection.

Scotch bonnet
mushroom

Marasmius oreades (Bolton : Fr.) Fr.

- ⊙ **Cap:** *up to 2 inches (5 cm) wide, often fluted or scalloped along the edge, reddish-ochre to ochre-beige, whitish as it dries.*
- ⊙ **Gills:** *emarginate, widely spaced, whitish then ochre-cream color; spore print white.*
- ⊙ **Stem:** *typically tough, slightly velvety, cream or reddish in color.*
- ⊙ **Flesh:** *thin, somewhat elastic; sweet flavor and pleasant odor described as cyanic, difficult to define.*
- ⊙ **Habitat:** *in fields and lawns, often in fairy rings.*
- ⊙ **When to harvest:** *Summer and autumn.*

Many mushrooms share the habitat of the Scotch bonnet (reported from North America and the UK) and some are poisonous. In order to avoid confusion, mycologists use the following method: after verifying that the mushroom possesses all of the Scotch bonnet features—color, spore color, emarginate and widely spaced gills—they hold the supposed Scotch bonnet in one hand at the base of the stem and, with the other hand, turn the cap in such a way that the stem winds around itself once or twice like a twisting rubber band. If the stem does not break and if the cap turns like a propeller when it is released, you probably have a Scotch bonnet. If it breaks, be careful!

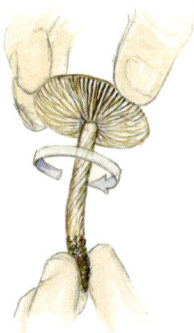

Warning

The toxic mushroom that is inadvertently harvested the most often when gathering Scotch bonnets is the fool's funnel (pp. 180–181). It grows in exactly the same places and sometimes even mixes in with the Scotch bonnet. Note that its cap is paler and often appears frosted over with a whitish bloom, its gills are distinctly more crowded, and its stem is brittle and does not pass the test mentioned earlier. *Marasmius collinus*, rare and also inedible (reported rarely from the UK, not from North America), strongly resembles the Scotch bonnet, but its stem is smooth, hollow, and fragile.

Golden bootleg

Phaeolepiota aurea (Matt. : Fr.) Konrad & Maubl.

- ⊙ **Cap:** up to 10 inches (25 cm) wide, powdery-floury texture all over, then wrinkled as it ages, beautiful ochre to bright tawny-ochre color.
- ⊙ **Gills:** emarginate, white, then ochre, and finally rust-colored; spore print ochre.
- ⊙ **Stem:** covered on the bottom by a "sock" called an armilla, or a "superior" annulus, powdery and floury texture like the cap and the same color, terminating at its summit in a large membranous ring.
- ⊙ **Flesh:** thick, whitish; sweet flavor and faint odor.
- ⊙ **Habitat:** in troops in grassy woods or along roadsides, usually in the mountains.
- ⊙ **When to harvest:** Summer and autumn.

This magnificent and large species is reported from northwest North America but not from the UK. It is easily recognized by its size, of course, but also and perhaps most of all by its cap and stem, which are both completely covered—at least in the youngest mushrooms—by a thick golden "flour" that tends to disappear in adults or in young mushrooms that have endured harsh weather. Its large ring is rust-colored on the top because of the mature spores that have fallen from the gills. Because it is a rare species, it should not be harvested; it is best to admire it in its home and protect it from being picked.

Warning

The golden bootleg is difficult to confuse with other mushrooms but does cause gastric poisoning in enough people to make it not recommended for the table. The big laughing gym (*Gymnopilus spectabilis*), reported from North America but not the UK, is similar to it in size. Its cap and stem, though, are not at all floury in texture and it grows in huge clumps on dead wood. Its very bitter flesh makes it inedible, which is a good thing because it is highly toxic.

Nameko

Pholiota nameko (T. Itô) S. Ito & S. Imai

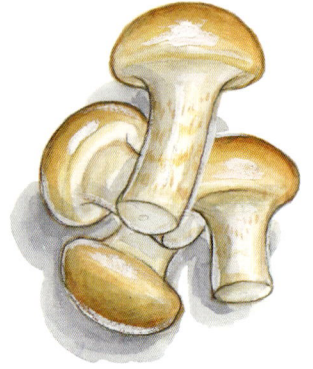

- ◉ **Cap:** up to 2.5 inches (6 cm) wide, viscous, sometimes with a few warts stuck within a gluey layer, bright ochre-yellow to reddish-yellow.
- ◉ **Gills:** emarginate, cream then rusty ochre color; rusty brown spore print.
- ◉ **Stem:** covered at the bottom by a "sock" or armilla, fibrillose, cream-colored, terminating in a fibrillose evanescent ring at the top.
- ◉ **Flesh:** whitish; sweet flavor and faint odor.
- ◉ **Habitat:** on dead wood from deciduous trees, sawdust shavings, etc.
- ◉ **Cultivated only.**

Nameko means "gelatinous mushroom" in Japanese. You have probably already eaten this mushroom without knowing it because it is one of the basic ingredients in the miso soup traditionally served in Japanese restaurants. Production of this *Pholiota* takes place primarily in Japan and China and it is not sold fresh in Europe. It can, however, be found in jars, either by itself or pickled in a "wild mushroom" mixture, sometimes mistakenly referred to as a "foliote."

Did you know?

The spores of the nameko contain allergens, and people sensitive to them who cultivate the mushrooms are sometimes stricken with a pulmonary disease called "mushroom lung," a pneumonia that presents itself as chronic coughing, fatigue, migraines, etc., all of which are symptoms linked to a loss of pulmonary capacity. It should be noted that other cultivated mushrooms (*Pleurotus*, shiitake, etc.) may also engender the same problems.

Gilded
brittlegill

Russula aurea Pers.

- ⊙ **Cap:** *up to 4 inches (10 cm) wide, dry and matte, bright orangey-red to orangey-yellow.*
- ⊙ **Gills:** *adnate, cream-colored with a lemon yellow edge; spore print ochre.*
- ⊙ **Stem:** *white or tinged with bright yellow or lemon yellow.*
- ⊙ **Flesh:** *crumbly, white; sweet flavor and faint odor.*
- ⊙ **Habitat:** *under deciduous trees.*
- ⊙ **When to harvest:** *Summer and autumn.*

This very pretty *Russula* mushroom (reported from the UK, but not North America) favors deciduous forests and is easy to identify. It loves warmth, which is why it is most often observed in the summer after heavy storms. Instead of examining the somewhat variable coloring of its cap to try and distinguish it from its many peers, it is more practical to look for the bright yellow edge of its gills and to examine its stem, which is sometimes tinged the same color. Its sweet flesh (see the following pages for more about the flavor of *Russula* flesh) makes it an excellent edible that is widely appreciated.

Did you know?

You may encounter a form of the gilded brittlegill that is very difficult to recognize because of its lack of yellow pigment and is quite uncommon. This form—called *axantha* from the privative prefix *a* and *xanthos* meaning yellow—can really only be identified using a microscope, for while it lacks the typical coloring it nevertheless possesses the same ensemble of microscopic traits as normally colored specimens.

- ⊙ **Cap:** *up to 4 to 6 inches (10–15 cm) wide, often a bit greasy to the touch, highly variable in color, usually a blend of gray, purple, pink, and green, but sometimes completely green.*
- ⊙ **Gills:** *adnate, whitish, pliable under the finger without breaking; if you rub them, it will feel as though you are running your finger over a piece of lard; spore print white.*
- ⊙ **Stem:** *white.*
- ⊙ **Flesh:** *brittle, white; sweet flavor and faint odor.*
- ⊙ **Habitat:** *under deciduous and coniferous trees.*
- ⊙ **When to harvest:** *Summer and autumn.*

Charcoal
burner

Russula cyanoxantha (Schaeff.) Fr.

The charcoal burner is a good quality edible mushroom, reported from the UK and North America, and frequently overlooked. While it comes in many colors, it can easily be recognized with one characteristic detail: though the gills of all other *Russula* break easily at the slightest pressure, those of the charcoal burner, soft and pulpy, bend beneath the finger without breaking. If they are vigorously rubbed, the sensation is one of touching something waxy and greasy, like lard. Gills of this kind exist in a few other closely related and edible *Russula* species as well. *Russula variata*, reported from North America but not the UK, differs because a bit chewed for some seconds (and spit out) is spicy. It is also edible, though not of the same quality.

In the kitchen

The charcoal burner has a pleasant hazelnut taste and is perfect for simple dishes. It should be harvested young because it quickly becomes filled with worms. It can be cooked in a pan and seasoned with salt,

The vast *Russula* group

Russula species are present all over the world, from fields in Arctic zones that are snowy much of the year to the stifling tropical and equatorial forests of Africa, South America, Asia, and Oceania. This very impressive level of diversity—there are more than 2,500 Russula described in the world today—has allowed Russula species to acquire a variety of forms that mycologists often find dumbfounding. Russula with rings like Amanita abound in Africa, for example, others look just like Lactarius species, and still others could pass for truffles.

pepper, and persillade (parsley sauce). Some people add a little bit of diced smoked bacon to this. To enjoy eating the charcoal burner, one must appreciate its firm flesh, because many edible *Russula* have rather crunchy flesh even after being cooked.

Did you know?

All *Russula* with greasy gills are edible; in addition to the charcoal burner, we can also mention *Russula cutefracta* (reported as a synonym of *cyanoxantha*), which looks uncannily like the quilted green russula (p. 97), and *Russula langei* (reported from the UK but not North America), a charcoal burner lookalike that has

very firm flesh and a stem whose surface is often flushed a lilac color. When it is completely bright green, the charcoal burner is said to be of the *peltereaui* variety.

Russet
brittlegill

Russula mustelina Fr.

- ◉ **Cap:** *up to 6 inches (15 cm) wide, very firm and even hard, ochre-brown to reddish-brown.*
- ◉ **Gills:** *adnate, white to cream; spore print yellowish to pale ochre.*
- ◉ **Stem:** *white then tinged with ochre.*
- ◉ **Flesh:** *very firm, white; sweet flavor and faint odor.*
- ◉ **Habitat:** *under coniferous trees, especially spruce trees, in mountainous areas.*
- ◉ **When to harvest:** *Summer and autumn.*

The russet brittlegill, reported from the UK and North America, is often the cause of much hand-wringing on the part of people hoping to harvest ceps: the color of its cap is indeed a perfect imitation of the "mushroom kings," and this *Russula* is often found overturned on the ground, showing its gills instead of the hoped-for tubes. Still, it is a good quality edible with firm flesh and a pleasant hazelnut flavor and is not easily confused with many other mushrooms. The nutty brittlegill (*R. integra*), reported from North America and the UK, may be all brown and grow in the same habitat, but its gills are yellow, and, in any event, it is edible. *Russula brunneola*, reported from North America, has a white spore print, but is also considered a good edible.

Did you know?

The only time a *Russula* species is considered inedible is if the flesh is too pungent or bitter, or stains red or black when bruised. For this reason, the flavor of the flesh is one of the first features used by mycologists to identify these mushrooms. In a nutshell, all of the *Russula* that are sweet to the taste are edible, but the true edibility of many is unknown, so caution is advised. An exception must nevertheless be made for the olive brittlegill (*Russula olivacea*, p. 203) (reported from the UK, and as edible in North America), which has been held responsible for a few fairly serious poisonings in countries where it is eaten in large amounts, northern Italy in particular.

Quilted green
russula

Russula virescens (Schaeff.) Fr.

- ⊙ **Cap:** *up to 6 inches (15 cm) wide, matte and crackled, dark bluish-green against a cream background.*
- ⊙ **Gills:** *adnate, crumbly, white to cream-colored; spore print white.*
- ⊙ **Stem:** *white then stained ochre at the base.*
- ⊙ **Flesh:** *white; sweet flavor and faint odor.*
- ⊙ **Habitat:** *under deciduous trees.*
- ⊙ **When to harvest:** *Summer and autumn.*

The quilted green russula, reported from North America and the UK, looks remarkably like *Russula cutefracta*, which fortunately is also edible: it has the same silhouette and the same crackled green cap. To tell the two species apart, just remember that *R. cutefracta*, reported as a synonym of *cyanoxantha*, is very close to the charcoal burner (p. 95) and has the same greasy gills. If you have a small amount of iron salts on hand (see p. 99), also note that the quilted green russula turns pink during the chemical reaction, while *R. cutefracta* does not react at all.

Did you know?

Other green *Russula* species exist, but none of them have a cap as crackled as those we have just described. The charcoal burner (p. 95) can be totally green in its *peltereaui* form, and you may also come across the greasy green brittlegill (*R. heterophylla*) with its forked and cross-veined gills around the stem, or the grass-green russula (*R. aeruginea*) that grows beneath birches and spruces. Both are reported from North America and the UK. In Mediterranean regions, *Russula monspe-liensis* only grows under sun roses, or cistus plants.

- **Cap:** up to 5 inches (12 cm) wide, "ham" pink, but also beige-pink or brown-pink, often faded in the center and dotted with tiny rust-colored spots; the skin of the cap is usually "too small" and the tips of the gills can be seen around the periphery.
- **Gills:** adnate, whitish, sometimes a little rust-colored, spore print white.
- **Stem:** whitish, often speckled with a rusty color at the base.
- **Flesh:** white; sweet flavor and faint odor.
- **Habitat:** under deciduous and coniferous trees.
- **When to harvest:** Summer and autumn.

Bare-toothed
russula

Russula vesca Fr.

Mycologists say that the bare-toothed russula, reported from North America and the UK, "shows its teeth" because the tips of its gills protrude beyond the edge of its cap. For the same reason, some people use an even stronger image and call it the "Russula in a miniskirt." While variable in color like many of its close relatives, the bare-toothed russula is easy to identify. It is a good edible with firm flesh but is nevertheless not widely known and is for the most part overlooked. It is sometimes very abundant in forests, usually in soil that is somewhat or not at all calcareous.

Warning

It should be noted that *Russula* and *Lactarius* are the only mushrooms with seedy flesh made up of miniscule grains, giving them a very unique consistency that is both firm and crumbly. It is often said that *Russula* stems break like pieces of chalk, which gives some idea as to the texture of their flesh. The *Lactarius* also possess this seedy flesh, but when they break they emit a juice that is more or less the consistency

of milk, which is what gives them their name.

Did you know?

In order to identify European *Russula*, mycologists always start by tasting them. This test poses no danger whatsoever if spit out and provides important information about the mushroom's identity depending on whether the flavor is sweet, tangy, or bitter. Next, the color of the spore print must be closely examined using color codes that appear in specialized publications. The bare-toothed russula has a white spore print, coded "Ia" by mycologists. The olive brittlegill (*R. olivacea*, p. 203) has

a rather bright yellow spore print coded "IVc." The microscope is also a useful tool, allowing mycologists to observe the spores and various elements that make up the skin of the cap, which is called the cuticle.

Iron salts

Iron salts exist in the form of greenish crystals. Mycologists interested in Russula *species always have a piece in their pocket because the colorful reactions it produces on* Russula *stems are good clues about their identity. If you rub this crystal on the stem of the charcoal burner, nothing will happen: the stem will remain white. On the other hand, if you perform the same test on a bare-toothed russula—or the russet brittlegill (p. 96)—a bright salmon pink reaction will occur. Iron salts are also helpful for telling apart the quilted green russula—a pink reaction—from* Russula cutefracta, *which does not react (see p. 97).*

Dark scaled
knight

Tricholoma atrosquamosum (Chevall.) Sacc.

- ◉ **Cap:** *up to 4 inches (10 cm) wide, covered with dark gray or black scales on a whitish background.*
- ◉ **Gills:** *emarginate, whitish or pale gray; spore print white.*
- ◉ **Stem:** *white, glossy, sometimes with a few fine black scales.*
- ◉ **Flesh:** *white; flour-like taste and sweet, floury odor when cut.*
- ◉ **Habitat:** *under coniferous trees.*
- ◉ **When to harvest:** *Autumn.*

The dark scaled knight, reported from North America and the UK, belongs to the "Petits-gris" or little grays, a group that includes almost every *Tricholoma* in the gray knight family (p. 106). They are good mushrooms for eating, rather easy to recognize, and grow abundantly. The dark scaled knight is often confused with *Tricholoma squarrulosum* (rarely reported from North America and not from the UK), also edible, which has a stem sprinkled with blackish scales and a cap covered with a thick wooly felt in younger specimens. The latter only grows under deciduous trees.

Warning

Watch out for the tiger tricholoma (p. 209), reported from North America but not the UK. It is highly poisonous and also has scales on its cap. It is substantially larger—its cap can easily surpass 6 inches (15 cm) in diameter—and its gills are usually a subtle shade of green, giving it its characteristic blue-green glimmer. Its flesh gives off a flour-like odor that tends to become quickly unpleasant. The ashen knight (*Tricholoma virgatum*), reported from North America and the UK, should also be avoided. It has a fibrillose cap and flesh that has a very peppery taste after being chewed for a few minutes and spit out. This is a harmless test that can be performed while you are harvesting.

Girdled
knight

Tricholoma cingulatum (Almfelt) Jacobashch

- **Cap:** up to 2.5 inches (6 cm) wide, felted, a more or less dark gray.
- **Gills:** emarginate, whitish or pale gray; spore print white.
- **Stem:** white or grayish with a white or ochre fleecy ring.
- **Flesh:** white; flour-like odor and flavor.
- **Habitat:** only under willow trees.
- **When to harvest:** Autumn.

In Europe, *Tricholoma* species with rings are rare, and within this small group the girdled knight can easily be recognized by its felted gray cap and its exclusive relationship with willow trees in both North America and the UK. The pleasant flour odor of its flesh is also a good clue. It is very close to the yellowing knight (p. 104), which is also edible but grows under other deciduous trees, does not have a ring on its stem, and has flesh that turns intensely yellow when it matures.

Did you know?

Tricholoma matsutake is another *Tricholoma* species with a ring that is very rare in most of Europe but well-known outside its borders. Its Latin name, *matsutake*, is taken directly from its Japanese name, meaning "mushroom of the pine trees," and describes its habitat perfectly. It is a beautiful mushroom covered with red scales that gives off a strong aromatic odor. It is highly regarded by the Japanese, who consume 1,000 tons of it each year, most of it imported from North America and northern European countries such as Sweden. In North America it is recognized as a separate species, *Tricholoma magnivelare*.

- **Cap:** up to 5 inches (12 cm) wide, a bit greasy to the touch at first, smooth and silky, peels easily forming triangular pieces.
- **Gills:** emarginate, white, irregular; spore print white.
- **Stem:** white, fibrillose, often marked at its base with blue spots.
- **Flesh:** white; sweet flavor and flour-like odor.
- **Habitat:** under deciduous trees.
- **When to harvest:** Autumn.

Blue spot
knight

Tricholoma columbetta (Fr. : Fr.) P. Kumm.

The less hairy *Tricholoma*

The name Tricholoma *comes from the Greek word* trichos, *meaning "fur, hair," that has given us many words in English like trichosis—any disorder involving hair— and trichotillomania, a condition in which a person has a constant urge to pull out his or her hair. And yet, there are no more hairs on a blue spot knight than there are on an egg! There are other* Tricholoma *species, however, that are clearly furry, like the scaly knight (*T. vaccinum*) (reported from North America and the UK, but not considered edible), which has a red cap covered with large fibrillose scales. The gray knight (p. 106) is also far from having a smooth cap, though it still cannot really be considered hairy.*

The blue spot knight, reported from North America and the UK, is not common and is sometimes difficult to distinguish from the other white *Tricholoma*: to avoid an incorrect identification, remember that its cap is silky and not matte, that the skin covering it peels easily, and that its flesh gives off a flour-like scent. Check that the spore print is white so that you are not confusing the blue spot knight with an *Entoloma*. The blue spots at the base of its stem are an excellent feature for recognition, but they are not always easy to see because they are often at the very bottom. If you don't want to miss out on this feature, harvest these mushrooms carefully and make sure the stem remains intact.

Warning

This beautiful edible mushroom is most likely to be confused with *Tricholoma* from the white knight group (p. 210), which are totally inedible and known for their matte cap— which tends to turn ochre in adult specimens—and the strong and often unpleasant odor of their flesh, a very aromatic scent that smells somewhat like gas. Also beware of possible confusion with deadly white *Amanita*, which emerge from a volva in a persistent sac at the base of the stem, have free gills, and a ring on the stem.

Did you know?

There are around 100 different *Tricholoma* species in Europe. Unlike some other kinds of mushrooms that often require a microscope for identification, almost all *Tricholoma* can be identified with the naked eye, provided you take the time to observe the ensemble of their characteristics and their habitat. Very few of them are poisonous, but many of them are inedible because they are too sharp or bitter in flavor.

Yellowing
knight

Tricholoma scalpturatum (Fr.) Quélet

- ◉ **Cap:** up to 3 inches (8 cm) wide, felted, silvery gray to dark gray, sometimes whitish, turns yellow as it ages.
- ◉ **Gills:** emarginate, whitish, yellowing in bruised areas and with age; spore print white.
- ◉ **Stem:** white or grayish, turns yellow.
- ◉ **Flesh:** white; distinctly flour-like flavor and odor.
- ◉ **Habitat:** under deciduous trees.
- ◉ **When to harvest:** Autumn.

This mushroom, reported from North America and the UK, is very common in woods made up of young trees, and is reported as a good edible in Europe, but is not considered edible in North America. It can be identified with just a little bit of valuable know-how. In order to recognize it, note that it only grows under deciduous trees and that its flesh turns yellow in mature specimens and gives off a straightforward scent of flour. The gray knight (p. 106), edible in both North America and the UK, only grows under fir trees (very rarely under deciduous trees), its cap is darker, its flesh does not turn yellow, and it is almost completely odorless.

The girdled knight (p. 101) strongly resembles the yellowing knight, but it only grows under willow trees and its stem has a downy whitish ring.

Warning

There are several other small gray *Tricholoma* that are not always easy to recognize. The tiger tricholoma (p. 209), an enormous species that grows near mountain conifers, has gills that are usually tinted green. The beech knight (*T. sciodes*) (a synonym for *T. terreum*), reported from North America and the UK, and the ashen knight, both of which are poisonous and grow beneath beeches and spruces respectively, have fibril-lose caps, gills edged in black, and a very disagreeable and sour-tasting flesh with a faint earthy or flour-like odor (for more details, see the gray knight, p. 106).

Coalman

Tricholoma portentosum (Fr. : Fr.) Quél.

- ⊙ **Cap:** up to 4 inches (10 cm) wide, a little greasy to the touch at first, fibrillose, gray to gray-black with a lighter edge.
- ⊙ **Gills:** emarginate, white with yellow tints; spore print white.
- ⊙ **Stem:** white, tinted with yellow.
- ⊙ **Flesh:** white; sweet odor and flour-like odor.
- ⊙ **Habitat:** under coniferous trees and some deciduous trees, most often in colder areas.
- ⊙ **When to harvest:** In autumn, sometimes late in season.

This *Tricholoma*, reported from North America and the UK, has a contradictory name: its scientific name comes from the Latin *portentosus*, meaning "marvelous, tremendous," but also "bizarre, monstrous." Let us assume that the naturalist who gave it its name, renowned Swedish mycologist Elias Magnus Fries, was trying to praise this beautiful mushroom rather than discredit it. Unfortunately, the hazardous history of common names had other plans. Either way, it is a lovely mushroom for eating and is quite common in certain areas.

Warning

The coalman may be confused with certain gray forms of the deceiving knight (*T. sejunctum*, or *subsejunctum*, as there is confusion over the correct name), reported from the UK and North America, which is inedible without being truly poisonous; its flesh is not sweet and a bit bitter or astringent, and its odor is only faintly flour-like. The base of the deceiving knight's stem is also typically pink. Among the other gray *Tricholoma* that grow beneath coniferous trees, beware of the ashen knight (*T. virgatum*), reported from North America and the UK, which is poisonous, never presents yellow hues, and has gills with a black border around the edge (for other gray *Tricholoma*, see the gray knight, p. 106).

Gray knight

Tricholoma terreum (Schaeff. : Fr.) P. Kumm

- ◉ **Cap:** *up to 3 inches (8 cm) wide, felted and wooly, generally dark gray.*
- ◉ **Gills:** *emarginate, whitish to pearl gray; spore print white.*
- ◉ **Stem:** *white or grayish.*
- ◉ **Flesh:** *white; faint flavor and odor, not flour-like.*
- ◉ **Habitat:** *under fir trees, more rarely under other conifers (very rarely under deciduous trees).*
- ◉ **When to harvest:** *Autumn.*

The gray knight, reported from North America and the UK, often forms impressive colonies under fir trees, and usually grows in the company of viscous boletes (*Suillus*, p. 139) that associate with the same tree species. It is a good edible mushroom that you simply have to know how to recognize: its cap is a rather dark gray and felted, its gills are delicately tinted gray, and its flesh is odorless. All of these features are critical to correctly identifying it. The dark scaled knight (p. 100) resembles the gray knight and also grows under conifers, but its cap bears visible scales and its flesh has an instantly sweet then flour-like smell when it is cut.

Warning

Like all edible gray *Tricholoma*, the gray knight may be confused with the tiger tricholoma (p. 209), a large mountain forest species with gills that are typically a greenish color. The ashen knight (*T. virgatum*) is toxic, grows under spruce trees, and has a fibrillose cap, gills with black edges, and a tangy flesh when chewed. *Tricholoma josserandii*, with very few reported occurrences worldwide, is also poisonous and also grows under fir trees, but its cap is greasy to the touch, often wrinkled, and its flesh gives off an odor of rancid flour. It is a good idea to pay close attention to these details.

Stubble
rosegill

Volvariella (Volvopluteus) gloiocephala (DC. : Fr.) Boekhout & Enderle

- ◉ **Cap:** *up to 6 inches (15 cm) wide, viscous at first then shiny as it dries, of variable color ranging from warm brown to pearl gray or white.*
- ◉ **Gills:** *free, white then pink, spore print pink.*
- ◉ **Stem:** *white or grayish, no ring but does have a large white membranous sac volva at its base.*
- ◉ **Flesh:** *white; sweet flavor and odor of radishes or raw potatoes.*
- ◉ **Habitat:** *grass in fields, meadows, and roadsides, sometimes also on soil receiving manure.*
- ◉ **When to harvest:** *Autumn.*

The stubble rosegill, reported from North America and the UK, is a common mushroom that feeds on decomposing grass. As a result, it is frequently seen in grassy areas when there is a high enough level of atmospheric humidity. It may be confused with ringless *Amanita* like the grisette (p. 54) which, thankfully, are not toxic. Their caps are not viscous but distinctly striated around the edge, and their gills are usually not pink except in adults and dried-out specimens.

Warning

Deadly *Amanita* species like *Amanita phalloides* (p. 229) can easily be confused with the stubble rosegill mushroom. Remember that *Amanita* always grow in forests or at least in close proximity to the trees they associate with, they have a ring around the stem that can sometimes fall off, that their spore print is white, not pink, and that their gills are bright white in adult specimens, never pink.

Did you know?

A species close to the stubble rosegill, the paddy straw mushroom (*V. volvacea*), can be found in a number of traditional Asian dishes. It is cultivated in rice paddies.

- ◉ **Cap:** *up to 8 inches (20 cm) wide, sometimes 12 inches (30 cm) in very developed specimens, beige to bluish-gray or gray-brown.*
- ◉ **Gills:** *decurrent, whitish or grayish; spore print white to lilac.*
- ◉ **Stem:** *very off-center and reduced in size, sometimes absent, whitish.*
- ◉ **Flesh:** *whitish; sweet flavor and pleasant mushroom odor.*
- ◉ **Habitat:** *in tufts on dead or living deciduous tree trunks.*
- ◉ **When to harvest:** *Autumn.*

Oyster mushroom

Pleurotus ostreatus (Jacq. : Fr.) P. Kumm

The oyster mushroom is by far the most common *Pleurotus* mushroom, reported from North America and the UK. It is sold in markets and grocery stores just about all year round and can be found in many pre-packaged forest blends. This mushroom's worldwide cultivation represents twenty-five percent of the entire cultivated mushroom market. It is encountered fairly frequently in nature and grows on tree trunks in cool and humid places, with a mild preference for trees with white and tender wood like willows, maple, and poplars.

Pleurotus at home

Many garden centers today sell culture kits for growing *Pleurotus* species at home. These kits include a bag of substrate seeded with the mushroom's mycelium (the white part). The bag must be perforated all over to allow the mushrooms to find their way out and then placed in a cool humid area. Several "series" of the mushroom crop will follow as long as the resulting mature specimens are regularly harvested. *Pleurotus* species can be cultivated without a kit just as easily: simply buy a few *Pleurotus* mushrooms and rub their gills on strips of white wood, like the wood used to make produce crates, for example. These strips can be placed in a cool spot in your garden, covered with a few handfuls of dirt, and kept humid. With a little luck, your omelette is not far behind!

Did you know?

Pleurotus have an excellent yield when they are cultivated and are capable of transforming 3.5 ounces (100 grams) of organic substrate into 1.75–2.5 ounces (50–70 grams) of fresh mushrooms containing up to thirty percent protein. In developing countries, *Pleurotus* mushroom domestication is an excellent way to recycle agricultural waste by using thatch left over from harvested grain as a substrate.

Pleurotus from here and elsewhere

While you might be lucky enough to find the oyster mushroom on a walk, you are less likely to stumble upon the pink oyster (P. salmoneostramineus) or the golden oyster mushroom (P. cornucopiae var. citrinopileatus) that are sometimes found in stores. Both of these species only grow in Asia. They are sometimes cultivated in Europe and North America, though only anecdotally in comparison with the quantities of the oyster mushroom produced.

King oyster
mushroom

Pleurotus eryngii (DC. : Fr.) Quél.

- ◉ **Cap:** up to 4 inches (10 cm) wide, somewhat sunken in the center, dark brown, reddish-brown or brown-gray.
- ◉ **Gills:** decurrent, spaced, white to cream; spore print white.
- ◉ **Stem:** centered or slightly off-center, whitish to cream.
- ◉ **Flesh:** somewhat tough, whitish; sweet flavor and rancid flour odor.
- ◉ **Habitat:** only on the roots of field eryngo (Eryngium), in dunes or lawns with calcareous soil.
- ◉ **When to harvest:** In autumn.

The king oyster mushroom has a very specialized habitat: it only grows when grafted onto the roots of a thistle from the *Asteraceae* family, called the eryngo or sea holly in English. There is a much larger variety (var. *ferulae*) that prefers carrot relatives from genus *Ferula* (*Apiaceae* family) instead. These *Pleurotus* species with a roughly central stem are often extremely difficult to identify when separated from their substrate, even for many mycologists. Perhaps because of this very specialized habitat, it has not yet been reported from North America or the UK though one day we may grow some of its hosts in our flower gardens!

Warning

Microscopic features are obviously very important in mushroom identification, and mycologists refer to them extensively. Among the mushrooms with white spore prints and decurrent gills, most *Pleurotus* can be recognized immediately by their somewhat cylindrical spores resembling small sausages. This characteristic is particularly useful if a harvester submits the king oyster for identification without having made a note of its specific habitat.

Cornucopia
mushroom

Pleurotus cornucopiae (Paulet) Rolland

- ⊙ **Cap:** *up to 6 inches (15 cm) wide, attached laterally to the substrate, sunken in the middle, cream to brownish-white or whitish as it dries.*
- ⊙ **Gills:** *very decurrent and typically forming ridges on the stem that are connected by irregular transversal partitions, white to cream; spore print white.*
- ⊙ **Stem:** *very off-center, covered by nets formed by the gills, white to cream.*
- ⊙ **Flesh:** *rather tough, whitish; sweet flavor and rancid flour odor.*
- ⊙ **Habitat:** *on dead wood from deciduous trees, often on trunks and logs.*
- ⊙ **When to harvest:** *Summer and autumn.*

This *Pleurotus* is fairly commonly reported from the UK and North America, on old fallen trees in cool areas. It is known for its pale hues, but also and most of all for its highly decurrent gills that form a kind of web decorating nearly the entire stem. Its rather strong odor is not always pleasant, especially when it gets older, so it should be consumed when it is young and fresh. You may confuse it with other *Pleurotus* species, but this would not put you in any danger. The pale oyster (*P. pulmonarius*), reported from North America and the UK, is just as edible, is whitish to pale gray-brown, but its gills are only slightly decurrent and its flesh gives off an aniseed odor.

Warning

Watch out for the lilac oysterling (*Panus conchatus*), an inedible mushroom reported from North America and the UK that has a silhouette similar to the cornucopia. Look for its distinct mauve or lilac tones, especially in young specimens and on the gills.

- ◉ **Cap:** up to 4 inches (10 cm) wide, somewhat sunken in the center, smooth, a uniform bright yellow.
- ◉ **Gills:** decurrent, folded, forked and connected by transversal veins, bright yellow like the cap; spore print pinkish-buff.
- ◉ **Stem:** whitish or yellow.
- ◉ **Flesh:** yellowish; sweet flavor and strong, very pleasant odor that is fruity and reminiscent of ripened apricots.
- ◉ **Habitat:** under various deciduous trees, on noncalcareous soil.
- ◉ **When to harvest:** Summer and autumn.

Chanterelle

Cantharellus cibarius Fr. : Fr.

Paradoxically, and contrary to popular belief, the "true" chanterelle that we are introducing here is by far the rarest in genus *Cantharellus*. Most recently mycologists are finding that many species, reported from across North America and the UK, have been called "cibarius," but differ from the "true" *cibarius*. Happily, all are edible. In the large majority of cases, the chanterelles that are sold commercially do not belong to this species and are in fact *Cantharellus pallens* (p. 115), which are quite commonly reported from northern Europe, though not the UK or North America. The latter species can be distinguished easily by its much paler colors and by its flesh, which is less scented and turns yellow. *Cantharellus ferruginascens*, reported from the UK, but not from North America, also turns quite yellow when touched; its cap is paler, less fleshy, and often has greenish or blue-green tones, at least in younger specimens.

Warning

Make sure not to confuse chanterelles with jack-o'-lanterns (p. 199), which are highly toxic but usually larger and more orange with well-formed gills. They grow for the most part in tufts at the base of trees, on logs or pieces of dead wood buried underground. The chanterelle is also often confused with the false chanterelle (p. 188): this mushroom grows on wood and possesses a soft and completely tasteless flesh that is fortunately not poisonous but is reported to cause gastric upset in some people.

Did you know?

Chanterelles are strange mushrooms. While most of their relatives create large quantities of spores in a very short period of time and die rapidly afterwards, chanterelles produce very few spores at a time, but they do so over a long period. Over the course of evolution, this unique mode of reproduction was eventually accompanied by another adaptation that allows chanterelles to live long enough to make their spores: they produce a variety of active substances that ward off insects, bacteria, and other fungi, allowing them to remain intact sometimes as long as three months.

Growing chanterelles: a challenge for researchers

As is the case with boletes, and for similar reasons (see p. 131), scientists have yet to master the art of cultivating chanterelles. The most a team of Swedish researchers managed to produce was a single chanterelle in a cultivation tray, and even that was only successful once. It is therefore not astonishing that the chanterelle is considered an endangered species in several countries in Europe.

Amethyst
chanterelle

Cantharellus amethysteus Quél.

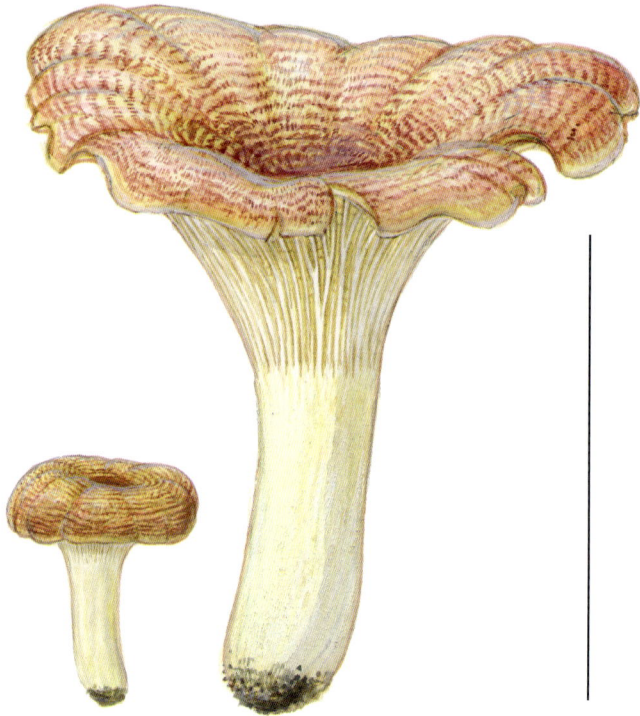

- ⊙ **Cap:** *up to 4 inches (10 cm) wide, somewhat sunken in the center, with small purplish scales on a dull yellow background, yellowing slowly when bruised.*
- ⊙ **Gills:** *decurrent, folded, forked and connected by transversal veins, ochre yellow.*
- ⊙ **Stem:** *ochre yellow, yellowing slowly but significantly when bruised.*
- ⊙ **Flesh:** *yellowish; sweet flavor and pleasant, fruity odor, spore print buff.*
- ⊙ **Habitat:** *under various deciduous trees, in noncalcareous soil.*
- ⊙ **When to harvest:** *Summer.*

Reported from across the UK, but not reliably from North America, the amethyst chanterelle is very easy to recognize: it is a summer chanterelle that turns bright yellow when damaged and whose cap is adorned with small purplish scales. These scales are at times few and far between—in which case a magnifying glass has to be used to distinguish them—or so abundant that the entire surface of the cap will be one color. When its cap is bare or just about, this species may be confused with *Cantharellus ferruginascens*, reported from the UK, but not from North America, which never has scales on its cap but is sometimes lilac in color. Its flesh yellows just as strongly as the amethyst chanterelle's.

Did you know?

To differentiate between species that are morphologically close, mycologists use a variety of characteristics that are only visible with the use of a microscope. In other mushroom groups these microscopic characteristics vary widely, but they are extremely uniform among chanterelles, which makes identifying them from one another a delicate task, even for specialists.

Pale chanterelle

Cantharellus pallens Pilát

- ◉ **Cap:** *up to 4 inches (10 cm) wide, somewhat sunken in the center, covered entirely at first by a fine white bloom that slowly disappears to reveal the yellow background.*
- ◉ **Gills:** *decurrent, folded, forked and connected by transversal veins, ochre-yellow.*
- ◉ **Stem:** *ochre yellow, yellows slowly but significantly when bruised.*
- ◉ **Flesh:** *yellowish; sweet flavor and a pleasant, fruity order that is not overpowering.*
- ◉ **Habitat:** *under various deciduous and coniferous trees, in noncalcareous soil.*
- ◉ **When to harvest:** *End of spring through autumn.*

This is the chanterelle that is sold by the crate throughout the markets of Europe. It is recognizable by its pale color and its yellowing flesh, which gives off a pleasant odor that is not too strong. But these differences are not enough; mycologists now consider *C. pallens* a synonym for *C. cibarius*. *Cantharellus ferruginascens*, sometimes bears a strong resemblance to it, but is less fleshy, more flaccid, and its cap is often a greenish or bluish green color, particularly around the edges. *Cantharellus atlanticus*, a recently described species, as yet reported only from France, has a cap that appears similarly frosted but grows under coastal fir trees and has bright orangey-yellow gills. *Cantharellus alborufescens*, now considered a synonym of *Cantharellus cibarius*, is sometimes found under Mediterranean oak trees. It is pale all over and yellows when bruised. "Chanterelles" are reported from across North America and the UK; one day the mycologists may differentiate between many more of the forms.

Did you know?

Though chanterelles are some of the best edible mushrooms, there is, puzzlingly, very little known about them, and mycologists are constantly discovering new species. *Cantharellus ilicis* was first described in Spain as recently as 2008; to date there are very few confirmed reports. It is a beautiful species that depends on the evergreen oak and has very pale gills. This new chanterelle has also been harvested under the same Mediterranean tree.

- ◉ **Cap:** *up to 4 and sometimes 6 inches (10–15 cm) wide, funnel-shaped, often with small black or blackish gray scales.*
- ◉ **Gills:** *absent; cinder gray to black under the smooth cap, spore print white to ochre.*
- ◉ **Stem:** *hollow inside the extension of the cap, blackish-gray.*
- ◉ **Flesh:** *thin, membranous; pleasant fruity flavor and odor.*
- ◉ **Habitat:** *under deciduous trees.*
- ◉ **When to harvest:** *late summer into autumn.*

Black trumpet

Craterellus cornucopioides (L. : Fr.) Pers.

The black trumpet mushroom needs no introduction. Reported from the UK and North America, it is called *Craterellus fallax* in eastern North America. In spite of its less than inviting French nickname—the trumpet of death, which thankfully refers to the fact that it grows around Toussaint, or All Saints' Day, and not to its edibility—this easily recognized mushroom is sold widely at the market. It can really only be confused with the ashen chanterelle (p. 118), which has distinct folds under its cap. Confusing these two is not dangerous because the ashen chanterelle is also edible. In addition to its perfumed flesh, one of the black trumpet's assets is that, like the chanterelle, it often grows in colonies, making it easy to harvest. Once again, though, you have to know how to spot it, because its dark coloring can blend in with the autumn colors on the forest floor.

In the kitchen

The black trumpet keeps very well when properly stored. After it has been washed thoroughly and had any plant debris removed from it, it can be hung on a thread in a cool dry room. A needle and sewing thread are all that is needed to create garlands perfect for drying. You can also purchase dehydrators, which are very efficient and contain a sieve as well as a heating system. Once a black trumpet is dry, it can be used as it is simply by rehydrating it in warm water or by pounding it into a powder to use as a condiment in soups, for example.

Did you know?

Though they were once mixed together for classification purposes, today the chanterelles and trumpets are distinguished from each other: there are the chanterelles (genus *Cantharellus*) that are rather fleshy with a cap that is sunken in the center but without an orifice connecting to the stem (never funnel-shaped, in other words), and there are the trumpets and other chanterelles (genus *Craterellus*) that are less fleshy, often membranous, and have a very sunken funnel-shaped cap with a navel that is connected with the stem in a tube formation.

Elias Magnus Fries, the father of mycology

Born in 1794 in Femsjö, Sweden, little Elias was interested in botany from a very early age. In 1810 when he was just sixteen, he published the sixth edition of his Flora of Femsjö (Flora Femsioensis). *It is mushrooms, however, not plants, that owe him the greatest debt, and even today it is impossible to be a true mycologist without owning at least one edition—for they have been reprinted several times—of his monumental works entitled* Systema mycologicum *(1821),* Elenchus Fungorum *(1828), and* Epicrisis systematis mycologici *(1836–1838). In the mycology world he is often compared to emminent botanist Carl Linnaeus, also Swedish, who left behind some 19,000 plant beds after his death.*

Ashen
chanterelle

Craterellus cinereus (Pers. : Fr.) Quél.

- ◉ **Cap:** *up to 2.5 inches (6 cm) wide, funnel-shaped, smooth or fibrillose, with a notched or fringed edge, black or blackish-gray.*
- ◉ **Gills:** *in the shape of folds, decurrent, ash gray; spore print white.*
- ◉ **Stem:** *an extension of the cap, blackish-gray.*
- ◉ **Flesh:** *thin, membranous; pleasant fruity flavor and odor.*
- ◉ **Habitat:** *under deciduous trees.*
- ◉ **When to harvest:** *Autumn.*

This chanterelle is often confused with the black trumpet (p. 117). It is rarely reported from North America, and not reported from the UK. These two mushrooms resemble each other a great deal and sometimes grow together, but it is possible to distinguish the ashen chanterelle by its smaller size and most of all by the distinct folds on the underside of its cap, which is smooth in the black trumpet. The same difference exists between the trumpet chanterelle (*Craterellus tubaeformis*, p. 120) and the yellow-leg chanterelle (next page); the former has folds and the latter has only a few irregular raised veins.

Did you know?

"Chanterelle" is an adaptation of the Latin word *Cantharellus*, which means . . . chanterelle, and comes from the Greek word *kantharos*, or "cup," referring to the unique shape of these mushrooms.

Yellow-leg
chanterelle

Craterellus lutescens (Pers. : Fr.) Fr.

- ◉ **Cap:** *up to 2.5 inches (6 cm) wide, funnel-shaped, smooth or fibrillose, brown to brown-yellow.*
- ◉ **Gills:** *absent; the underside of the cap is smooth or marked by a few large irregular veins that are yellow to orangey-yellow or grayish; spore print pale orange to ochre.*
- ◉ **Stem:** *hollow inside the extension of the cap, yellow to orangey-yellow or grayish.*
- ◉ **Flesh:** *thin, membranous; pleasant fruity flavor and odor.*
- ◉ **Habitat:** *under coniferous trees.*
- ◉ **When to harvest:** *Late summer until the beginning of winter.*

Though it comes in a variety of colors, the yellow-leg chanterelle, reported from across the UK and North America, is easily identified by its funnel silhouette and the underside of its cap, which is either smooth or carpeted with large and irregular veins. It looks a lot like the trumpet chanterelle (*Craterellus tubaeformis*, p. 120) and grows in the same areas, but the latter has folds and is usually darker in color. Like the black trumpet, these chanterelles are well-suited to dehydration and, as long as they are not attacked by insects, can be stored this way for several months. Be cautious when eating it for the first time as it causes gastric upset in some people. When harvesting young specimens, try to avoid any confusion with the jelly baby (*Leotia lubrica*), reported from North America and the UK. See *Craterellus tubaeformis*, p. 120.

Did you know?

The yellow-leg chanterelle is a favorite of large mammals. In fact, it is not rare to see entire colonies of this mushroom razed to the ground by red deer and the European roe deer. Astonishingly, these animals do not seem to enjoy *Craterellus tubaeformis*, perhaps because it is less fragrant.

Trumpet
chanterelle

Craterellus tubaeformis (Bull. : Fr.) Quélet.

- ◉ **Cap:** *up to 2.5 inches (6 cm) wide, funnel-shaped, fibrillose or covered with small scales, brown, brown-gray to brown-yellow, sometimes completely yellow.*
- ◉ **Gills:** *in the form of distinct folds, decurrent, gray-yellow to gray-brown; spore print cream to yellow.*
- ◉ **Stem:** *hollow in the extension of the cap, ochre yellow to yellow-brown.*
- ◉ **Flesh:** *thin, membranous; pleasant fruity flavor and odor.*
- ◉ **Habitat:** *under coniferous trees.*
- ◉ **When to harvest:** *Late summer until the beginning of winter.*

The trumpet chanterelle is easy to recognize. It is another very fragrant mushroom that often grows in colonies across the UK and North America, and can be stored after drying, so it would be a great shame to be unable to identify it. But use caution when first trying this mushroom as it may cause gastric distress.

Warning

Very young specimens of the trumpet chanterelle and the yellow-leg chanterelle (p. 119) can sometimes be confused with the jelly baby (*Leotia lubrica*), a small yellow mushroom reported from North America and the UK that is shaped like a large gelatinous screw that grows in the same areas. It contains large quantities of toxic hydrocarbons, and it is best to avoid consuming it accidentally.

Did you know?

Mycologists generally agree that in the evolutionary history of mushrooms, the first kind to appear were ones with simple forms and caps with smooth undersides. Mushrooms with folds beneath the cap came next, increasing the surface area available to form spores and augmenting reproductive efficiency. True gills developed last and still exist today in a great number of mushrooms.

The fact remains, however—because evolutionary science is never simple—that some mushrooms with folds never moved on to the more "evolved" gill version. This may be the case for the chanterelles and trumpets, which would make them vestiges of very ancient morphological mushroom types.

Pig's ear

Gomphus clavatus (Pers.) S. F. Gray

- **Cap:** can reach 4 inches (10 cm) wide, fleshy, at first a bright purple then darkens little by little to an ochre-brown with age.
- **Gills:** in the form of very irregular folds, decurrent, the same color as the cap; spore print ochre to brownish.
- **Stem:** hollow inside the extension of the cap, same coloring.
- **Flesh:** thick, whitish; pleasant mushroom flavor and odor.
- **Habitat:** in mountain coniferous forests.
- **When to harvest:** Summer and autumn.

The pig's ear is well-known to mountaineering mushroom enthusiasts in warmer climates. It often grows in colonies across North America and the UK and cannot be mistaken for anything else: its colors, spinning top silhouette, and habitat are unique. Unfortunately, a victim of its own success, it is becoming rare and even starting to disappear from certain areas, to such an extent that some have proposed adding it to the European Red List of species threatened with extinction at the European level. It is therefore important to protect its status by not harvesting it; its future probably depends on the decisions we make now.

Did you know?

Though it is known in France as the "chanterelle violette," or violet chanterelle, the pig's ear has nothing in common with other chanterelles apart from the folds that cover the inferior face of its cap, which are simply the result of morphological convergence, something observed frequently in the natural world. Think, for instance, of dragonfly wings and those of flying fish. In reality, the pig's ear is a close relative of certain mushrooms from genus *Ramaria* (pp. 153–154).

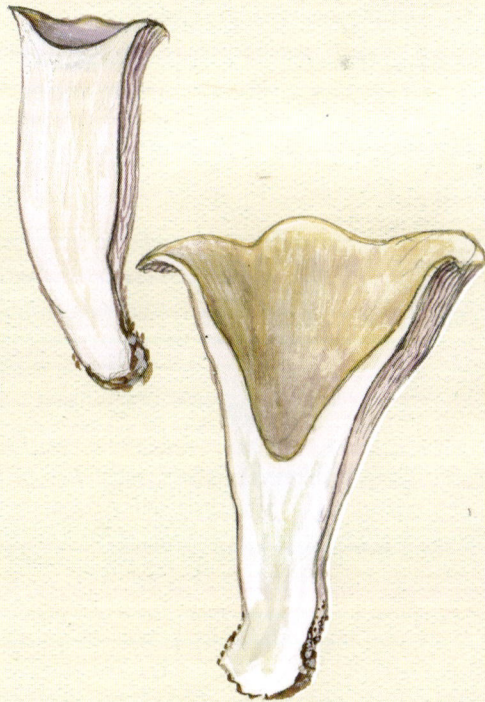

Hedgehog

Hydnum repandum (Scop. : Fr.) Fr.

- ◉ **Cap:** can reach 4 to 6 inches (10–15 cm) in width, fleshy, matte, cream to ochre orange.
- ◉ **Teeth:** decurrent, fairly crowded, cream; spore print white.
- ◉ **Stem:** firm, white then ochre.
- ◉ **Flesh:** firm but very brittle, white; sweet or somewhat bitter flavor and faint odor.
- ◉ **Habitat:** under deciduous and coniferous trees.
- ◉ **When to harvest:** Late summer through autumn.

The hedgehog, reported from North America and the UK, is very well-known and widely harvested. It is unmistakeable—or almost, because there are other similar-looking *Hydnum* species (all of which are edible, fortunately). The terracotta hedgehog (next page), for instance, is a miniature and slenderer version of the hedgehog, with teeth that are further apart and adnate, and flesh that slowly reddens when bruised, particularly on the surface of the stem. The less common *Hydnum albidum*, reported from North America but not the UK, has a white cap and stem, and only the teeth possess an ochre cream or apricot color. Other mushrooms like the *Hydnellum* have teeth under their caps, but they are usually darker and leatherier, which makes most of them inedible.

In the kitchen

If you want to sautée up a batch of hedgehog mushrooms, it is best to harvest young and fresh specimens only. Adult and aging specimens tend to become bitter. For the same reason, freezing the hedgehog is not advised. It is not, truth be told, a mushroom that keeps very well at all.

Terracotta
hedgehog

Hydrum rufescens Pers. : Fr.

- ◉ **Cap:** up to 1.5 inches (4 cm) wide, convex, orange to red-orange.
- ◉ **Teeth:** somewhat decurrent, rather crowded, cream or orangey-cream.
- ◉ **Stem:** slender, firm, cream and turns red when bruised.
- ◉ **Flesh:** firm but very brittle, white; sweet or somewhat bitter flavor and faint odor.
- ◉ **Habitat:** under deciduous and coniferous trees.
- ◉ **When to harvest:** Autumn.

The terracotta hedgehog is usually harvested under the shorter name of "hedgehog" because the morphological differences between the two species are not noticeable to most people. In fact, even mycologists now think the name *Hydnum rufescens* is a synonym for *Hydnum repandum*. It is not yet reported from North America or the UK. Still, this little mushroom with decurrent teeth and reddening flesh is easily recognizable. It is often more bitter than its older brother, and therefore deserves to be classified separately. It is easily confused with the depressed hedgehog, *Hydnum umbilicatum*, reported from North America but not the UK, which has the same silhouette and just about the same coloring, but its cap is heavily sunken in the center.

Did you know?

Over the course of mushroom evolution, teeth appeared independently within groups that were not directly related to each other. While the morphological convergences between the various groups that acquired teeth are obvious to the naked eye, one need only examine the various species under a microscope to detect their considerable differences: the spores of the *Hydnum* are smooth and colorless, while those of the *Hydnellum*, for example, are often brown and covered with spiny or warty protuberances; they are also leatherier.

- **Cap:** *up to 10 inches (25 cm) wide, dry, usually a beautiful dark and warm brown or blackish-brown, sometimes discolored as ochre-brown.*
- **Tubes:** *thin and crowded, cream then yellow and finally olive-yellow; spore print olive-brown.*
- **Stem:** *fleshy and often ventricose, brownish-beige or less dark, with a fine web at the summit that is often difficult to see.*
- **Flesh:** *firm, white; sweet flavor and very pleasant odor.*
- **Habitat:** *under deciduous trees, especially oaks.*
- **When to harvest:** *Especially in summer and early autumn.*

Bronze
bolete

Boletus aereus Bull. : Fr.

Enormous bronze boletes in the tropics?

European mushroom lovers on vacation or living as expats in the tropics often tell tales of memorable "heart-stopping" moments after stumbling upon enormous ceps strongly resembling bronze boletes that are largely overlooked by the local population. Unfortunately, these mushrooms are actually inedible or indigestible boletes belonging to genus Phlebopus. *In Australia, where these boletes are common, some mushrooms achieve impressive proportions, with caps one yard in diameter and a weight of nearly seventy pounds. That is certainly large enough to cause trouble for more than a few unknowing enthusiasts.*

All ceps are boletes, but not all boletes are ceps! There are four boletes in Europe, known as "noble" boletes, and they are often mistaken for one another and referred to generically as ceps: in addition to the bronze bolete, reported from the UK but rarely reported from North America, the other mushrooms awarded this superior gastronomic appellation are the king bolete (p. 129), the summer bolete (p. 127), and, rarest of all, the pine bolete (p. 131). The bronze bolete is distinguished by its dry and typically dark brown cap, the underdeveloped white web covering its colorful stem, and finally by its season: as soon as the middle of autumn arrives, it is rarely ever encountered.

In the kitchen

The bronze bolete can be cooked like other ceps, and this book alone would never be long enough to describe every recipe that pays tribute to its firm and fragrant flesh. It should be consumed when the mushrooms are young, though, and when the tubes have not yet turned an olive green; its flesh becomes soft with age and can also become wormy.

Did you know?

The undersides of bolete caps are lined with tubes encasing their pores, but this is not a defining feature because other mushrooms share this characteristic: certain polypores (from the Greek *poly*, "many" and *poros*, "pore") look almost exactly like boletes, but their flesh, instead of being soft and at times spongy, is often particularly tough, with a consistency like leather, cork, or wood, making them basically inedible.

- **Cap:** up to 8 inches (20 cm) wide, dry and often subtly velvety, uniformly light buff brown to ochre brown, more rarely warm chestnut brown (see next page).
- **Tubes:** thin and crowded, cream then yellow and finally an olive-yellow color; spore print olive-brown.
- **Stem:** fleshy and often ventricose, brownish-beige, with a clearly formed and obvious web often extending to the base.
- **Flesh:** firm, white even just below the surface of the cuticle; sweet flavor and very pleasant odor.
- **Habitat:** under deciduous trees, especially oaks.
- **When to harvest:** From the very beginning of summer to early autumn.

Summer bolete

Boletus reticulatus Schaeff.

The summer bolete is well-known but very often incorrectly identified. Commonly reported from the UK, there are only rare reports of the summer bolete from North America. It is frequently confused with the king bolete (p. 129) even though they are quite different if one pays attention. The king bolete's cap is usually glossy like lard and becomes lighter and lighter in color approaching the edge, which often contains a fine whitish zone, its web is much less developed, and the flesh beneath the skin is a light brown color (visible when cut). As its name indicates, the summer bolete is also the earliest of the ceps and appears beginning in late spring, provided it is sufficiently warm and rainy.

Warning

The summer bolete is easy to mistake for the bitter bolete (p. 221), which, though it is not poisonous, has such an unpleasant flavor that it makes cocktail bitters taste like a cool sip of water. Instead of ruining a plate full of mushrooms, just remember that the bitter bolete grows under conifers, that its tubes become pink with age, and that its web is more pronounced than the summer bolete's. If you are still unsure, taste a small piece of the raw cap and spit it out: this harmless test will help you avoid more unfortunate surprises.

Did you know?

Two very distinct summer bolete colorings can be observed in nature: one with a relatively pale ochre brown cap, the other with a warm chestnut brown or blackish-brown cap reminiscent of the bronze bolete. While mycologists have for a long time considered them to be simply two ecological variations within the same species, recent studies seem to show that the darker form is actually a separate species that is closer to the king bolete than the summer bolete. It will take a few more years of research to confirm this and to name this new bolete.

Many more unknown

Even though ceps, for obvious culinary reasons, are some of the most well-known mushrooms on the surface of the earth, mycologists are still discovering new species, even in Europe. In 2009, Swedish mycologists discovered a mushroom growing under pine trees that looked very much like the summer bolete but was not the pine bolete (p. 131). A new Latin name had to be invented for this mushroom, and it was baptized Boletus pinetorum. It remains rarely collected and has yet to be reported from the UK or North America.

- ◉ **Cap:** up to 10 inches (25 cm) wide (sometimes more), typically glossy and a little greasy to the touch, varying from whitish to dark brown, lighter and lighter in color approaching the edge.
- ◉ **Tubes:** thin and crowded, cream then yellow and finally olive-yellow; spore print olive-brown.
- ◉ **Stem:** fleshy and often ventricose, rather pale then ochre, a more or less pronounced web that is often limited to the top of the stem.
- ◉ **Flesh:** firm, white, a little brownish-pink just beneath the cap's skin; sweet flavor and very pleasant odor.
- ◉ **Habitat:** under deciduous trees and coniferous trees.
- ◉ **When to harvest:** Spring or autumn (rare in summer).

King
bolete

Boletus edulis Bull. : Fr.

The king bolete, reported from across North America and the UK, is often confused with the summer bolete (p. 127), which luckily is also edible. The king bolete's smooth and sometimes greasy cap, which loses color progressively approaching the edge, is nevertheless its classic feature. It is highly variable in color, and young specimens are sometimes completely white. There has been a significant amount of squabbling over the supposed differences in flavor linked to a king bolete's habitat: a number of people

would say growing under conifers is best, while others insist that it does not deserve to go in the pan unless it grows under deciduous trees. In this area it is difficult to be objective. Mycologists continue their struggle to determine the true species in this complex. For now, in eastern North America, it is called *B. chippewanensis.*

In the kitchen

This bolete can be prepared in many ways: in a creamy sauce, in a salad (still always cooked), grilled, in a soufflé, and even in a dessert, caramelized with vanilla custard. To store it, choose specimens that are young and firm with tubes that are still whitish or yellow. It stores well dried, frozen, or pickled. It is better

not to overwash it and it should never be soaked, because this might take away some of the flavor. Removing impurities with a knife and wiping it down with a clean towel is usually enough to get it ready for cooking.

Warning

The bitter bolete (p. 221) strongly resembles the king bolete. It is not poisonous, but its unbearable bitterness may leave you with no other choice but to throw a whole dish into the trash. Knowing how to identify it, therefore, is very important: its tubes rapidly turn pink, it has a rather brownish-pink spore print, its stem has a visible and pronounced dark web, and it grows under conifers. If you are not certain, taste a small piece.

- ⊙ **Cap:** *up to 10 inches (25 cm) wide, often wrinkled or dented, red-brown that is more or less dark, often covered with a fine white bloom in young mushrooms that is visible around the edge of the cap.*
- ⊙ **Tubes:** *thin and crowded, white then yellow and finally an olive-yellow; spore print dark olive.*
- ⊙ **Stem:** *fleshy and often ventricose, white then ochre, with a fine web.*
- ⊙ **Flesh:** *firm, whitish; sweet flavor and very pleasant odor.*
- ⊙ **Habitat:** *especially under fir trees, but sometimes under spruces as well as beeches, oaks, and chestnuts.*
- ⊙ **When to harvest:** *Autumn.*

Pine
bolete

Boletus pinophilus Pilát & Dermek

This magnificent cep is reported from the UK, and only rarely from North America, but it is nevertheless quite common across Central Europe. Whether it is growing under conifers, as is most often the case, or deciduous trees, it is easy to recognize its warm colors and the white bloom visible on younger specimens. It is not the only cep that grows under conifers, and the king bolete and the recently described *B. pinetorum* (see p. 127) share its habitat. In markets, pine boletes are often seen mixed in with large crates of king boletes.

Warning

If you buy ceps off the shelf, choose mushrooms that are young and free of any trace of mold. This may seem obvious, but like certain fruits, ceps should only be bought when they are in season. Mushrooms that have been kept for months in cold rooms— and treated with who knows what to have them ready for sale during the holiday season—cause poisonings every year.

Did you know?

The name "pine bolete" is confusing to some mushroom enthusiasts be- cause it makes them think of other boletes that associate with fir trees like the slippery jack (p. 139) or the granulated bolete (p. 138). The French-derived word *cep*, mean- ing "trunk," originates from the Latin word *ciptus*, or "column," probably referring to these boletes' massive stems. Its qualitative adjective *pinophilus* is formed from the Latin *pinus*, "fir tree," and the Greek *philos*, "friend," which describe this bolete's affinity for these conifers.

When will we be able to grow our own ceps?

Not anytime soon, unfortunately! Specialists today are able to cultivate around 100 mushroom species, 30 of which are produced at an industrial level, but all of them are saprotrophs (see p. 28). The biology of the mycorrhiza (the intimate relationship between the mushroom mycelium and the roots of larger plants) has not yet revealed all of its many secrets, and for this reason mycorrhizal mushrooms remain difficult to cultivate. As luck would have it, the large majority of excellent edible mushrooms, including ceps, fall into this category.

Butter
bolete

Boletus (Butyriboletus) appendiculatus Fr.

- ◉ **Cap:** *up to 8 inches (20 cm) wide, dry, chestnut to reddish-brown or russet brown.*
- ◉ **Tubes:** *thin and crowded, pale citrine yellow then greenish-yellow as the mushroom ages, turn slightly but distinctly blue when bruised; spore print olive-brown.*
- ◉ **Stem:** *fleshy, grows thinner at the base, yellowish stained with red at the bottom, with a fine yellow web at the top.*
- ◉ **Flesh:** *firm, whitish or yellowish; sweet flavor and very pleasant odor.*
- ◉ **Habitat:** *under deciduous trees, especially oaks, typically in noncalcareous soil.*
- ◉ **When to harvest:** *Summer into early autumn.*

This beautiful bolete, reported from the UK and rarely from western North America, grows in the middle of summer when most of the other edible mushrooms have given themselves over to the heat. It is easily confused with the iodine bolete (*B. impolitus*) (reported from the UK but rarely from North America), which is an inedible mushroom whose flesh releases an iodine smell at the base of its stem and does not have a ring. The butter bolete's flesh turns blue, if only faintly, and its cap is most often a brown that is distinctly reddish or russet. It does not like calcareous soil and prefers the company of oak trees and beeches. A lookalike, *Boletus subappendiculatus*, not reported from North America and rarely from the UK, grows in the mountains and associates with spruces. Its flesh does not turn blue at all.

Warning

There are other boletes that grow in the summer and they are not all edible. The devil's bolete (p. 219) has flesh that turns blue and is all too often consumed in the place of ceps or other well-known boletes. However, its pale cap and fat stem, typically a bright pink at the base, are enough to set it apart. See also the pretender (p. 215).

Scarletina bolete

Boletus (Neboletus) erythropus Pers.

- ⊙ **Cap:** up to 8 inches (20 cm) wide, dry and somewhat velvety, a beautiful dark to blackish-brown.
- ⊙ **Tubes:** thin and crowded, yellow, leading to orangey-red pores, all of which turns a bright blue; olive-brown spore print.
- ⊙ **Stem:** yellow to orangey-yellow covered all over with bright red dots, no network pattern (even at the top of the stem).
- ⊙ **Flesh:** yellow, turns very blue; sweet flavor and pleasant odor.
- ⊙ **Habitat:** under deciduous trees and coniferous trees, most often in noncalcareous soil.
- ⊙ **When to harvest:** Autumn.

Reported across North America and the UK, this is a very easy bolete to identify, and a good quality edible in spite of the way its flesh turns rapidly and intensely blue. It is considered edible, with caution, in the UK, and is reported to cause gastro-intestinal problems in North America. It is often rejected when mistaken for the devil's bolete (p. 219), though these two mushrooms have almost nothing in common. The deceiving bolete (p. 134), on the other hand, strongly resembles the scarletina bolete because of its stem—which bears no trace of a network pattern—but its cap is usually not as brown, and its flesh is usually beet red at the base of the stem. Any confusion between these two species is harmless because the deceiving bolete is just as edible, though of lesser quality.

Warning

There is another bolete without a network pattern that may be confused with the scarletina bolete: *Boletus lupinus* (p. 213) is distinctly larger, its cap is pinker, and its flesh turns only slightly blue. The lurid bolete (p. 214) has a network pattern on its stem and a red line just above the tubes that is visible when the cap is cut. In general, very few boletes that stain blue and/or have red pore mouths are considered edible, and then only for very experienced foragers who are able to observe subtle differences between those that are, and those that are not, edible.

Deceiving bolete

Boletus (Suillellus) queletii Schulzer

- ⊙ **Cap:** up to 6 inches (15 cm) wide, dry and somewhat velvety, varies in color, brown, orangey-brown to coppery red-brown.
- ⊙ **Tubes:** thin and crowded, olive-yellow leading to yellow pores that turn orangey-red, all of which turns a bright blue; spore print dark olive-brown.
- ⊙ **Stem:** yellow to orangey-yellow, typically beet red at the base, no network pattern (even at the top of the stem).
- ⊙ **Flesh:** yellowish and beet red at the base, turns bright blue; sweet flavor and pleasant odor.
- ⊙ **Habitat:** under deciduous trees, oaks in particular.
- ⊙ **When to harvest:** Summer and autumn.

The deceiving bolete is very commonly reported in the UK, but not in North America, and is closely related to the scarletina bolete (p. 133), with which it shares not only blue-staining flesh, but also a stem without a network pattern. It can be easily identified by the beet red color on both the inside and outside of the base of its stem. Only one other bolete, *Xerocomus (Xerocomellus) dryophilus*, reported from western North America but not from the UK, shares this characteristic, but it is much smaller and rarer. Confusing the two is not dangerous because the latter is not at all poisonous. The lurid bolete (p. 214) is another potential trap because its stem is sometimes completely devoid of any network pattern. In this case, Bataille's line (see p. 214) is a helpful feature for differentiating these two species. See also *Boletus lupinus* (p. 213).

Warning

There are some instances in which blue-staining boletes do not turn blue at all! Victims of either genetic anomalies or soil deficient in certain elements, these abnormal boletes can confuse even the most experienced mycologists. As stated previously, in general, very few boletes that stain blue and/or have red pore mouths are considered edible, and then only for very experienced foragers who are able to observe subtle differences between those that are and those that are not edible.

B a y
bolete

Boletus (Picipes) badius (Fr. : Fr.) Fr.

- ⊙ **Cap:** *up to 6 inches (15 cm) wide, a bit viscous at first then dry, russet brown to ochre-yellow.*
- ⊙ **Tubes:** *whitish then olive-yellow, turn blue when touched; spore print olive-brown.*
- ⊙ **Stem:** *smooth, ochre to reddish-brown.*
- ⊙ **Flesh:** *whitish, turns slightly blue; sweet flavor and faint odor.*
- ⊙ **Habitat:** *especially under conifer trees, very rarely under deciduous trees.*
- ⊙ **When to harvest:** *Summer and autumn.*

The bay bolete, reported from the UK and across North America, often stumps mushroom lovers because it has a cap with coloring that resembles the king bolete (p. 129) but its tubes turn blue! Nevertheless, it is a good quality edible that is worth knowing how to recognize because it is often overlooked for the aforementioned reason. If you can put together the brown cap, the blue-staining tubes, and the brown stem without a network pattern, it is hard to confuse with any other mushroom. Certain forms with very short tubes that do not turn blue are sometimes called *B. badiorufus* (re- ported from the UK, not from North America) but many mycologists believe that these are simply abnormal bay bolete specimens.

Warning

The bay bolete has an unfortunate habit of accumulating any and all pollutants that come within its reach and it has a special affinity for radioactive elements, which is why it was used to measure the progression of the radioactive cloud after the Chernobyl disaster. It is best not to harvest it in areas that are clearly polluted—roadsides, urban forests, etc.—or in regions where there is significant natural radioactivity.

Orange
bolete

Leccinum albostipitatum Den Bakker & Noordel

- ⊙ **Cap:** up to 8 inches (20 cm) wide, convex, fleshy, bright orange or orangey-brown.
- ⊙ **Tubes:** white then gray-brown; spore print olive-brown.
- ⊙ **Stem:** long, covered in fine scales that remain white for a long time before turning reddish or blackish; base of stem is sometimes blue-green.
- ⊙ **Flesh:** thick, white turning lilac-gray then blackish-gray when cut; sweet flavor and faint odor.
- ⊙ **Habitat:** only under poplars, especially aspens.
- ⊙ **When to harvest:** Summer and autumn.

It would be better to talk about the orange *boletes*, plural, because this common name covers several species that are mistaken for each other on a regular basis. These boletes are sometimes called "rugged" mushrooms—in honor of their coarse and unpolished-looking stems—or pibles, meaning "masts," as a result of which some are referred to as orange masts and others as brown masts. That being said, the true orange bolete, rarely reported from the UK and not from North America, can be identified by its stem covered in scales that remain white for some time, and by its unwavering affection for the poplars it depends on to grow. The rugged boletes—*Leccinum* species, to mycologists—are all edible (with caution in North America as poisonings have been reported from undetermined species of orange capped *Leccinums*), but those with an orange cap seem to be the most enjoyable to eat.

Did you know?

Other *Leccinum* species resemble the orange bolete a great deal. The orange oak bolete (see next page) causes the most confusion: connected to oak trees as the name indicates, in its youth the mushroom stem is covered in russet and then blackish scales, which should be enough to tell it apart from other mushrooms. The orange birch bolete (*L. versipelle*), reported from across North America, where it has been reported to cause gastric distress, and the UK, has dark brown or brown-black scales when young, and only grows under birch trees.

Orange
oak bolete

Leccinum aurantiacum (Bull.) S. F. Gray

- ◉ **Cap:** *up to 4 inches (10 cm) wide, convex, fleshy, brick red to orangey-brown.*
- ◉ **Tubes:** *white then gray-brown; spore print brown.*
- ◉ **Stem:** *long, covered in fine scales that are russet-colored even in young specimens and become blackish with age.*
- ◉ **Flesh:** *thick, white turning purplish-gray then black-ish-gray when cut; sweet flavor and faint odor.*
- ◉ **Habitat:** *various deciduous trees, often under oak trees.*
- ◉ **When to harvest:** *Summer and autumn.*

Like the orange bolete (see previous page), the orange oak bolete is one of the rugged boletes characterized by their orange caps and long stems covered with colorful scales. While most of them associate with only one kind of tree, despite its name the orange oak bolete has multiple relationships and can be found not only under oaks but also under poplars, birches, and even beeches, chestnuts, and willow trees. It can be identified primarily by the russet-colored scales that decorate its stem, which are visible even in very young specimens.

Did you know?

Just a few years ago, the orange oak bolete had the Latin name *Leccinum quercinum*—from *quercus,* "oak"— and the orange bolete was called *Leccinum aurantiacum,* the current name given to the orange oak bolete. Mycologists realized that the name *L. aurantiacum* had been applied to the wrong species for decades when they noted that the name's creator had been describing a bolete with russet-colored scales. The orange bolete was rebaptized as *L. albostipitatum* from *albus,* "white," and *stipes,* "stem."

Granulated
bolete

Suillus granulatus (L.) Roussel.

- ⊙ **Cap:** up to 5 inches (12 cm) wide, viscous, red-brown to ochre yellow.
- ⊙ **Tubes:** light yellow to greenish-yellow, exuding small drops of whitish milk in young specimens; spore print yellow-brown.
- ⊙ **Stem:** white or yellowish, studded at the top with small white or red granules.
- ⊙ **Flesh:** white in the cap, yellow in the stem; sweet flavor and faint odor.
- ⊙ **Habitat:** only under conifer trees.
- ⊙ **When to harvest:** Summer and autumn.

The granulated bolete is probably the most common of the "viscous boletes" in genus *Suillus*, reported from across North America and the UK. It is known for its exclusive relationship with conifer trees, its fairly dark reddish-brown cap, and its ringless stem, which is flecked near the top with small granules that are white at first and then a russet color when the mushroom is mature. When the mushroom is young and atmospheric conditions permit, the tubes exude small drops of whitish milk. It is most often confused with *Suillus collinitus*, reported from the UK but not North America, also edible, which has a browner cap and a distinctly pink mycelium, and the bovine bolete (*S. bovinus*), reported from the UK but rarely from North America (there perhaps associated with Scot's Pine), a mediocre edible whose flesh blues slightly, and is smaller, softer, and possesses many angular pores that tend to be arranged radially in mature mushrooms.

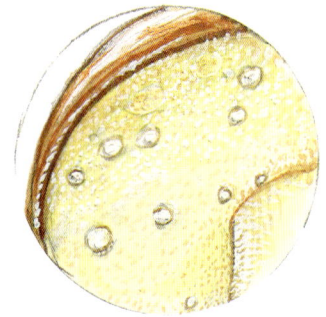

In the kitchen

The granulated bolete is a proper edible mushroom that often grows in large quantities. Take care not to harvest it from roadsides: like all mushrooms, it tends to accumulate heavy metals and other pollutants. The viscous film on the cap has mild laxative properties and should be carefully peeled off with the tip of a knife. Grilled and seasoned with a persillade (parsley sauce), this little bolete is quite pleasant to eat even if it may not have the same lofty qualities as its brothers, the "noble" boletes (p. 125).

Slippery **jack**

Suillus luteus (L. : Fr.) Roussel

- ◉ **Cap:** *up to 5 inches (12 cm) in diameter, viscous, light brown to reddish-brown.*
- ◉ **Tubes:** *pale yellow to greenish-yellow; fine pores; spore print ochre-brown.*
- ◉ **Stem:** *white or yellowish, with a wide white or purplish-brown membranous ring.*
- ◉ **Flesh:** *cream to yellowish; sweet flavor and faint odor.*
- ◉ **Habitat:** *only under conifer trees.*
- ◉ **When to harvest:** *Summer and autumn.*

The slippery jack mushroom, reported from across North America and the UK, is the only viscous bolete growing under conifer trees that has fine pores and a ring, so it is very easy to identify. Like other edible boletes with viscous caps, its cap should be peeled before eating because the gluey layer has laxative properties. It is the most commonly used bolete in commercial forest blends and is often referred to simply as a "bolete." While somewhat less common than the granulated bolete (p. 138), the slippery jack is nevertheless found quite frequently, especially on grassy and mossy trails in fir tree forests.

Did you know?

There are a number of other boletes in genus *Suillus*, and all of them are dependent on a single kind of tree with which they form an exclusive association. The larch bolete (*S. grevillei*), reported from North America and the UK, has a beautiful orangey-yellow cap and only grows under larch trees, just like the hollow bolete (*S. cavipes*), reported from North America and the UK, which has a felted reddish-brown cap that is not viscous at all. The slippery white bolete, *Suillus placidus*, reported from North America but rarely from the UK, prefers the company of pine trees with bundles of five needles like the Weymouth pine or the white pine. This bolete–tree specificity is obviously very important for identification.

Red
cracking bolete

Xerocomus (Xerocomellus) chrysenteron (Fr. : Fr.) Fr.

- ⊙ **Cap:** up to 5 inches (12 cm) wide, dry, with a tendency to crack in every direction when mature, dark brown to dark brown-gray or ochre-tobacco brown.
- ⊙ **Tubes:** dirty yellow then olive-colored, sometimes turn blue; spore print olive-brown.
- ⊙ **Stem:** smooth, often reddish and yellowish-burgundy color at the top.
- ⊙ **Flesh:** yellowish in the cap and reddish-burgundy in the stem; sweet flavor and faint odor.
- ⊙ **Habitat:** especially under conifers, rarely under deciduous trees (beeches).
- ⊙ **When to harvest:** Summer and early autumn.

It might look innocent, but this little bolete, reported from North America and the United Kingdom, is quite the trickster and is rather difficult to recognize. Perhaps because of the difficulty of identification several sources do not recommend it as edible in North America. Several closely related species bear a strong resemblance to it and one would have to be an experienced mycologist to be able to tell them apart. Fortunately, none of them are poisonous. The matte bolete (*X. pruinatus*), not yet reported from North America or the UK, for example, is often mistaken for the red cracking bolete. It appears later, from autumn through early winter, its flesh is more yellow, and it grows indiscriminately beneath various deciduous and coniferous trees. The brown forms of the ruby bolete (see next page) have flesh at the very base of the stem marked with blood red pinpricks.

Did you know?

Genus *Xerocomus* has only been studied in a systematic fashion for a few years. These studies have brought to light some new generic divisions and several species that had previously gone unnoticed, and several more will be described in the years to come. The revelation of these new species is not so astonishing when we consider that, according to the most optimistic mycologists, only five percent of mushrooms present on Earth today have been discovered.

Ruby
bolete

Xerocomus (Hortiboletus) rubellus (Krombh.) Quél.

- ⊙ **Cap:** *up to 4 inches (10 cm) wide, dry, highly variable in color, bright raspberry red to dark brown.*
- ⊙ **Tubes:** *yellow then olive-colored, turn more or less blue; spore print olive-brown.*
- ⊙ **Stem:** *smooth, yellow or reddish.*
- ⊙ **Flesh:** *pale yellow, typically with tiny blood red dots like insect bites at the very bottom of the stem, turns only slightly blue; sweet flavor and faint odor.*
- ⊙ **Habitat:** *under various deciduous trees.*
- ⊙ **When to harvest:** *Summer and autumn.*

The ruby bolete, reported from the UK and North America, is probably one of the most common *Xerocomus* species. At least in North America, it is not recommended as edible. Unfortunately, it is very rarely ruby-colored! It is more often a mixture of brown and pink, and in these brown forms that it is sometimes called *Xerocomus communis*, now recognized as a synonym of *Xerocomellus chrysenteron*, from the Latin *communis* meaning "common" or "banal," a name that is perhaps more suitable. Oddly enough, this name apparently has several species hiding behind it that are often confused with one another. *Xerocomus (Hortiboletus) bubalinus* (reported from the UK but not from North America), which grows in parks and urban forests and associates with poplars, is very difficult to distinguish from other mushrooms were it not for the fact that the flesh in its cap is usually pink.

Did you know?

Genus *Xerocomus* was created to group together small boletes with dry caps, a cylindrical stem, and irregular tubes shaped more like polygons than cylinders that tear apart when separated from one another. In genus *Boletus*, the tubes are cylindrical and separate from each other easily without tearing into half-tubes.

Suede
bolete

Xerocomus subtomentosus (L. : Fr.) Quél.

- ◉ **Cap:** *up to 4 inches (10 cm) wide, dry and felted, highly variable in color, from red to bright lemon yellow.*
- ◉ **Tubes:** *bright yellow gold in young specimens, then an olive-yellow, somewhat blue-staining; spore print olive-brown.*
- ◉ **Stem:** *smooth or possessing pronounced veins, pale or bright yellow; white or pale yellowish mycelium.*
- ◉ **Flesh:** *pale or whitish-yellow, usually somewhat pink in the lower half of the stem; sweet flavor and faint odor.*
- ◉ **Habitat:** *under various deciduous trees.*
- ◉ **When to harvest:** *Summer and autumn.*

Like almost all other *Xerocomus* species, the suede bolete is a model of variability, and many enthusiasts find it hard using one name for a bolete that can be red, brown, or bright yellow! The tubes and pores are usually bright yellow, however, and the pinkish flesh with the pale mycelium are helpful clues. Reported from across North America and the UK, it is reported edible in the UK, but not so in North America as it is easily confused with other, little known species.

It may be confused with *Xerocomus (Boletus) ferrugineus*, reported rarely from North America, but frequently from the UK, which has a bright yellow mycelium and white flesh.

In the kitchen

None of the *Xerocomus* are poisonous, but their soft and creamy flesh is not usually well-liked. At the very least, they can help round out a meager harvest of better-quality edibles.

Did you know?

Two other small boletes have recently been described that are close to the suede bolete and *Xerocomus ferrugineus* and have bright yellow mycelia: *X. silwoodensis*, not reported from the UK or North America, grows under poplars, its cap is red, and the top of its stem has protruding veins; *X. chrysonemus* strongly resembles the suede bolete, but it has bright yellow flesh in the base of its stem and grows under oak trees; it is rarely reported from the UK but not from North America.

Beefsteak
mushroom

Fistulina heptica (L.) P. Kumm.

- ◉ **Cap:** *up to 6–8 inches (15–20 cm) wide, shaped like a tongue attached to dead wood; upper face is gelatinous (when humid) with papillae, salmon pink or reddish-pink.*
- ◉ **Tubes:** *fine, not fused (when examined with a magnifying glass), reddish-pink; spore print yellow-red.*
- ◉ **Stem:** *absent or short, fleshy.*
- ◉ **Flesh:** *reddish-pink with lighter striations; tart flavor and faint odor.*
- ◉ **Habitat:** *often at the base of trees or on stumps, favors oak and chestnut trees.*
- ◉ **When to harvest:** *Late summer into autumn.*

Mycologists use the name "polypore" to categorize mushrooms that are typically without a stem, grow on the wood of tree trunks, branches, or stumps, and have a cap whose inferior face is carpeted with tubes terminating in small holes (the pores). Usually extremely leathery, very few of the mushrooms in this group are edible. The beefsteak, reported from across the UK and North America, is a notable and especially interesting exception because it is an edible mushroom that provokes strong reactions: while its savory and acidic flesh is not for everyone, it certainly deserves to be tasted.

In the kitchen

The beefsteak can be cooked in a variety of ways depending on your tastes and imagination, but its unique flavor does not pair well with everything. Here is a simple recipe that highlights its flavor: cut the mushroom in half-inch slices, then bread and sauté them in a tablespoon of butter, seasoned with salt and pepper.

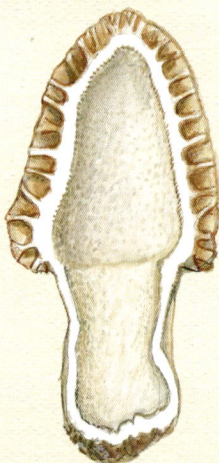

- ⊙ **Cap:** *up to 2.75 inches (7 cm) wide and 2.75 inches (7 cm) tall, often globular (but not always) and made up of deep, uneven pits that are beige, yellowish-brown, and sometimes tinted red.*
- ⊙ **Stem:** *completely hollow, cream, just about smooth; spore print ochre.*
- ⊙ **Flesh:** *thin, somewhat fragile; sweet flavor and faint odor.*
- ⊙ **Habitat:** *under deciduous trees in cool rich soil in sparsely wooded areas, parks, and gardens, often beneath ash, elm, tulip, oak, poplar, sycamore, and apple trees.*
- ⊙ **When to harvest:** *Only in the spring, except early summer in high mountains and colder regions.*

Yellow morel

Morchella esculenta (L. : Fr.) Pers.

Growing morels . . . just a dream?

Cultivating morels has been a project undertaken by many teams of scientific researchers in recent decades, but morels are fickle and do not allow themselves to be easily tamed. Their biology is poorly understood, their vegetal partners—if they exist—are not well-known, and their habit of appearing in one location one year only to never grow there again remains puzzling. Nevertheless, several teams around the world are apparently in the process of finalizing a method for mass-producing morels, and a few websites even offer farming tunnels for this highly prized mushroom. Let's see what the future has in store . . .

It is always a treat to stumble upon the beautiful yellow morel, perhaps while walking along a river under ash trees just beginning to reveal their first leaves to the springtime sun. Its beige cap is often round and composed of deep and irregularly shaped pits that are not aligned with one another, and its stem is cream-colored and hollow, making it an easy mushroom to recognize. The *crassipes* variety stands out for its greater height and a disproportionately thick stem that can reach up to 5 inches (12 cm) wide. This mushroom is reported from across North America and the UK, and mycologists are studying their diverse genetics, and finding many new species. For us, it will be enough to divide them into yellow morels and black morels.

Warning

Morels must be fully cooked to become non-toxic, and even then some people cannot eat them without gastric distress. They should never be eaten raw. Morels are most likely to be confused with false morel species (p. 248), which can be deadly when consumed under certain circumstances. The false morel's cap is made up of fat folds resembling the grooves in the cerebral cortex instead of pits and its stem is not entirely hollow. This mushroom can often be confused with an inedible *Peziza* mushroom called the bonfire cauliflower, which grows in areas that have previously burned and can reach an imposing size: up to 10 inches (25 cm) tall! It is easily distinguished from morels by its lack of a stem and is more similar in appearance to a *Sparassis* (p. 283).

Are morels rare?

We are not all equal in the eyes of morels. While some regions are overflowing with them, other areas have almost none at all, which many enthusiasts view as a true injustice. In some regions, morels can be expensive, (1) because they are not farmed, but harvested naturally, and (2) because of their rarity.

Black
morel

Morchella elata (Vent.) Pers.

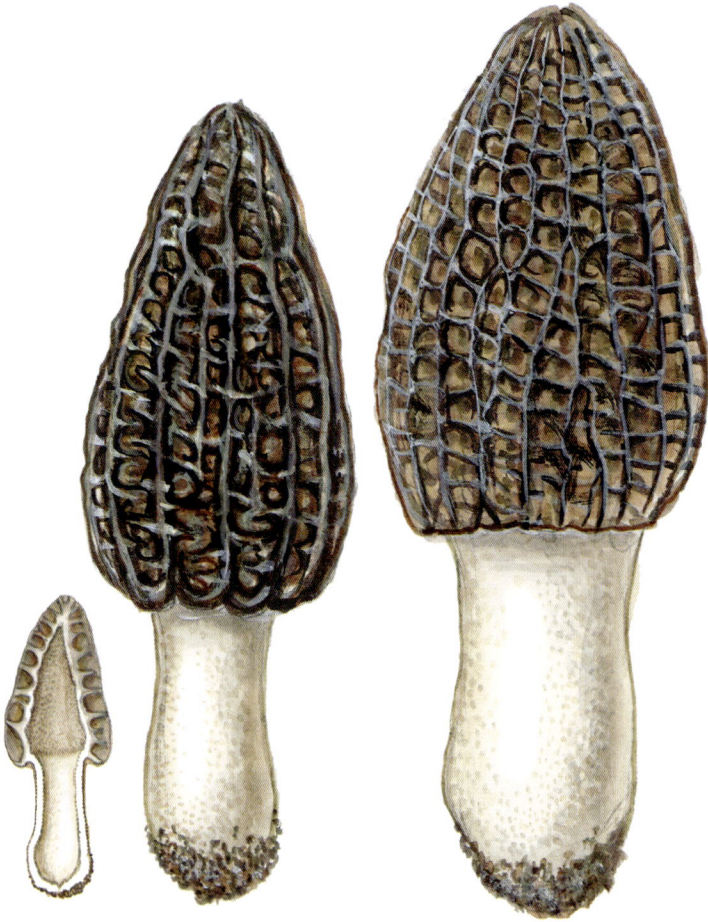

- ⊙ **Cap:** up to 2.5 inches (6 cm) tall and 1.25 inches (3 cm) wide, shaped like a cone set atop a stem and made up of irregular brown pits (alveoli) that extend vertically in roughly parallel grooves; the edge of the cap is connected to the stem by a depression called a vallecula.
- ⊙ **Stem:** completely hollow, cream, finely granular.
- ⊙ **Flesh:** thin, somewhat fragile; sweet flavor and faint odor.
- ⊙ **Habitat:** under deciduous and coniferous trees in cool rich soil in sparsely wooded areas, parks, and gardens.
- ⊙ **When to harvest:** Only in the spring, except early summer in high mountains and cold regions.

Morels are traditionally separated into two groups: the "brown" and the "blonde." The black morel is at the head of the line in the first group, while the yellow morel (p. 145), as its name suggests, features in the second. Morels are often confused with the early morel (p. 148), which is also edible and has a cap made up of crude folds that rests on top of the stem like a thimble. The black morel is reported from across North America and the UK, fruiting somewhat earlier than the yellow morel.

Warning

Do not confuse the black morel with the false morel (p. 248). Cook morels thoroughly, never eat them raw.

Did you know?

Morel morphology is highly variable. Everything—the shape of the cap, its color, size, the arrangement and number of alveoli, its habitat—is subject to variation. For many years, mycologists considered all of these different ecological forms to be separate species, and morel identification was very difficult. Recent studies based on comparisons of small DNA fragments seem to suggest that all of these forms are in fact simply variations of a few highly polymorphic species.

Half-free morel

Morchella (Mitrophora) semilibera DC.

- ◉ **Cap:** *up to 1.5 inches (4 cm) tall and 1.25 inches (3 cm) wide, in the shape of a cone on top of a stem and made up of irregular brown alveoli.*
- ◉ **Stem:** *completely hollow, cream, velvety, with between one-third and one-half of its length penetrating into the cap.*
- ◉ **Flesh:** *thin, somewhat fragile; sweet flavor and faint odor.*
- ◉ **Habitat:** *under deciduous trees in cool rich soil in sparsely wooded areas, parks, and gardens.*
- ◉ **When to harvest:** *Only in the spring, except early summer in high mountains and colder regions.*

Reported from North America, and the UK, nothing looks more like a morel than a half-free morel. This mushroom grows in the same areas as its older sisters and at the same time of year. It can be found in the cool and dark parts of sparsely wooded areas, and often grows in parks, gardens, and even riparian forests. It is the only "morel" whose stem is not fused to the edge of its cap, and instead the stem pushes up into the cap between one-third and one-half of its height. It should be eaten cooked because, like morels, it is mildly toxic if consumed raw.

Did you know?

The fact that the half-free morel's cap is not fused to its stem like other morels has always intrigued mycologists, and it was for this reason that the half-free morel was set apart for a time in genus *Mitrophora*. Today's mycologists know that this characteristic is not reason enough to separate it from the true morels. However, the mycologists have found several different species of the half-free morel; those in the UK may not be the same species as those in North America, and those on the east and west coasts of North America may not be the same species.

Early
morel

Verpa bohemica (Khrombh.) J. Schröt.

- ◉ **Cap:** up to 2.5 inches (6 cm) tall and 1.25 inches (3 cm) wide, shaped like a thimble and resting on top of the stem, traversed by large irregular wrinkles.
- ◉ **Stem:** hollow, cream, finely granular.
- ◉ **Flesh:** thin, somewhat fragile; sweet flavor and faint odor similar to that of semen in adult specimens.
- ◉ **Habitat:** under deciduous trees in cool rich soil in sparsely wooded areas, parks, and gardens.
- ◉ **When to harvest:** Only in the spring, except early summer in high mountains and colder regions.

Reported from North America, but not from the UK, like the half-free morel (p. 147), the early morel looks very much like a true morel and generally grows in the spring. It is easily recognized by its cap, which rests on the stem like a thimble and is very wrinkled instead of pitted with alveoli. It may be confused with another edible mushroom, *V. digitaliformis* (reported from the UK but not North America) whose cap is similar in shape but not very wrinkled. Like morels, it should be eaten cooked because it is mildly toxic when raw.

The *Helvella* mushrooms (next page) grow more in autumn, their cap is made of membranous lobes, and their stem is covered with thick ribbing.

Warning

The greatest risk of confusion comes from the early morel's resemblance to a small false morel (p. 248): it should be noted that the latter grows beneath conifers and that its cap is usually darker with the edge connected to the stem; it does not at all resemble a thimble. Also watch out for stinkhorn (p. 299) specimens that have lost the green and smelly substance covering their caps.

White
saddle

Helvella crispa (Scop. : Fr.) Fr.

- ⊙ **Cap:** *up to 2.5 inches (6 cm) wide, made up of irregular and fragile membranous lobes, whitish or cream-colored on the upper face, pale ochre and felted on the lower face.*
- ⊙ **Gills:** *absent.*
- ⊙ **Stem:** *whitish, completely hollow and creased.*
- ⊙ **Flesh:** *thin and membranous, somewhat elastic but rather brittle; faint flavor and odor.*
- ⊙ **Habitat:** *under deciduous trees.*
- ⊙ **When to harvest:** *Autumn.*

The *Helvella* species are quite common mushrooms, reported from North America and the UK, and easy to identify: their stem is often marked by deep veins that make it appear creased, and their cap consists of a few large and irregular membranous lobes. The white saddle is happiest in and around sparsely wooded areas with deciduous trees and can also be found in parks and gardens. The fluted black helvella or slate gray saddle (*H. lacunosa*), also reported from across North America and the UK, resembles the white saddle, but is much darker in color: its cap is blackish-gray and its stem is gray to brownish-gray. It is just as edible as its fuzzier cousin, but not as popular.

Did you know?

Like their cousins the morels (pp. 145–146), *Helvella* species are mildly toxic when raw. They are generally not recommended as edible in North America or the UK as they may contain the same toxins as *Gyromitra*, which can accumulate, later to cause death. If you choose to eat them, they should always be cooked well, and any water used in cooking should be discarded. They contain hemolytic substances that destroy red blood cells, but these are quickly destroyed by heat or dehydration. Many *Helvella* in stores are sold already dehydrated.

Orange
peel

Aleuria aurantia (DC. : Fr.) Maire

- ◉ **Mushroom in cup shape:** *rests directly on bare ground, may reach 4 inches (10 cm) wide.*
- ◉ **Flesh:** *thin, brittle, somewhat paler; sweet flavor and faint odor.*
- ◉ **Habitat:** *on exposed earth, often along the edge of trails.*
- ◉ **When to harvest:** *Sometimes in spring and summer, but most of all in autumn.*

Finding this pretty *Peziza* is always a welcome surprise, though more for aesthetic reasons than anything else: while it is not poisonous, it is particularly bland. It lends itself best to culinary fancies, and could be used raw to decorate a salad, for example. Some people also eat it as an appetizer, candied in kirsch. It is reported from across North America and the UK. It may be confused with another brightly colored *Peziza*, the ruby elfcup (*Sarcoscypha coccinea*), reported from North America and the UK, which is not poisonous, though not considered edible, and grows grafted onto dead wood by a small white stem in winter and spring. It is smaller and redder than the orange peel.

Did you know?

In mushrooms that have a cap, spores generally form on the cap's inferior face, whether it is smooth or covered with gills, pores, tubes, or teeth. In *Peziza* species, though, the fertile area is on the inside of the cup. If you harvest an adult orange peel, place it under a lamp that gives off a little heat while the mushroom is still fresh and pay close attention, you will probably see billows of "smoke" formed by millions of microscopic spores as they escape from the center of the cup. Each spore, shaped like a rugby ball, measures about fourteen thousandths of a millimeter in length.

▲ *Ruby elfcup*

Bleach
cup

Disciotis venosa (Pers.) Boud.

- ◉ **Mushroom initially cup-shaped,** then flattens out on the ground and is able to reach 6 inches (15 cm) in width, light brown to dark brown superior face with large and prominent wrinkles in adults, light brown to ochre inferior face.
- ◉ **Flesh:** brittle, not very thick; sweet flavor and distinct bleach odor.
- ◉ **Habitat:** on the ground in areas that are cool and rich in humus, under deciduous trees.
- ◉ **When to harvest:** Only in the spring.

Now here is a large *Peziza* reported from northeastern North America and the UK that is easy to identify: it grows at the same time and in the same areas as morels (pp. 145–146), it is colored various shades of brown, its internal face has prominent wrinkles, and it gives off an odor that sometimes tickles the nose. It is known for being an edible of excellent quality, and the odor quickly disappears during cooking. It favors rich and humid soil, is happiest along streams, and will not shun large city parks when the weather is cool enough.

Did you know?

The thick cup (p. 222) looks a great deal like the bleach cup and also grows in the springtime, but it possesses the same level of toxicity as the false morel (p. 248) in so it is best to avoid confusing the two species. The thick cup is a redder brown color and grows under conifers, on mossy logs, or near old campfire sites in mountainous areas. In addition, its flesh does not have a characteristic odor. It may also be confused with *Gyromitra leucoxantha*, reported rarely from North America and the UK, which has a yellow cup, and *Discina ancilis*, reported from across North America and the UK, which does not smell of bleach. Pay attention to these details.

Judas ear,
Tree ear

Auricularia auricula-judae (Bull.) Quélet

- ⊙ **Mushroom in irregular cup shape:** *attached in the center to dead wood; the external face is delicately velvet, and the internal face is often pleated, creating an overall shape that sometimes resembles a human ear.*
- ⊙ **Flesh:** *elastic when fresh, very tough when dry; sweet flavor and faint odor.*
- ⊙ **Habitat:** *on dead or living wood, often on elder trees in Europe, but on conifers in North America.*
- ⊙ **When to harvest:** *Just about all year long, apart from periods of extreme cold.*

This very common mushroom, reported from North America and the UK, is recognized as a separate species in North America: *Auricularia americana*. It is easy to recognize: its appearance and the elastic texture of its flesh are its classical features. It cannot be said that this mushroom is very tasty, and it is enjoyed for its consistency more than anything else: it is the "black mushroom" in Asian cuisine. It keeps very well when dried, and only needs to be left in warm water for a few minutes to rehydrate and be ready to use. Other *Auricularia* are eaten under the same generic black mushroom label and sold in Europe, including *A. polytricha*, a tropical species with a velvety external face covered in a whitish felt. These "ears" are very easy to cultivate on dead wood by sowing pieces of the mycelium on logs, for example, and then storing the seeded logs in a place with constant and elevated humidity (between fifty and eighty percent) and an average temperature above 68°F. In North America, the Judas ear or tree ear may be confused with *Tremella foliacea*, which grows in foliose clumps, and *Excida recisa*, which grows on deciduous wood.

Did you know?

According to the Bible, Judas Iscariot hung himself on an elder tree, which seems unlikely considering this shrub's stature and the fragility of its wood. Nevertheless, it is because of this story that this mushroom, which looks like an ear and often grows on elders, was given this strange common name. Medical research seems to show that under certain conditions it may have anticoagulative, hypoglycemic, and hypocholesterolemic properties. There are reports that eating great quantities may interfere with blood clotting.

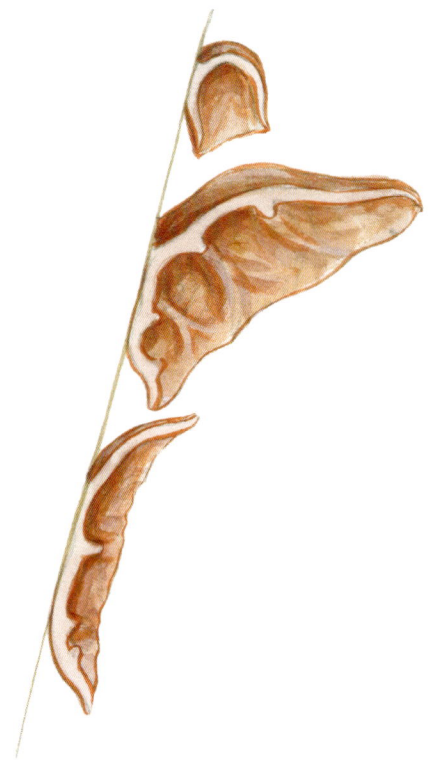

Golden
coral

Ramaria aurea (Schaeff.) Quél.

- ⊙ **Mushroom in shrub shape:** *branched, up to 4 inches (10 cm) tall and 2.75 inches (7 cm) wide, lemon yellow to light orangey-yellow, ochre-yellow to pale salmon-yellow when mature; spore print brownish-yellow.*
- ⊙ **Stem:** *thick and often short, fleshy, whitish to lemon yellow.*
- ⊙ **Flesh:** *white, thick; sweet flavor and pleasant odor.*
- ⊙ **Habitat:** *on the ground beneath deciduous trees, especially beeches in Europe, under conifers in North America.*
- ⊙ **When to harvest:** *Summer and autumn.*

The golden coral is reported from eastern North America and the UK. Yellow and orange *Ramaria* are difficult to distinguish from each other, and the golden coral is no exception; it is often necessary to use a microscope in order to identify it with certainty. You should consider eating the golden coral only if you have confidently identified it with a microscope. However, it is not recommended as edible in North America, and the difference in habitat may be significant. It is still a good edible with a pleasant taste that is easy to harvest because it often grows in small colonies. The golden coral is most likely to be confused with other close species that may be harmless though perhaps less tasty. *Ramaria largentii* (reported rarely from western North America and not from the UK), for example, is a more intense orangey-yellow color and grows only under conifers, in particular fir trees and spruces; its edibility is not known. It may also be confused with *Ramaria flava*, which is not considered edible, and is reported from the UK and North America. *Ramaria formosa*, reported from across North America and the UK, is similar to the golden coral, and is considered toxic in North America.

Warning

The upright coral (p. 223) has fairly significant laxative properties and may be confused with the golden coral. It always grows in close proximity to dead wood and is usually sticking to branches on the ground, its stem forms long white mycelium threads, and its flesh is fairly bitter and astringent. See also the salmon coral (p. 225).

Clustered
coral

Ramaria botrytis (Pers. : Fr.) Bourdot

- ⊙ **Mushroom in shrub shape:** *branched, reaching up to 6 inches (15 cm) tall and 8 inches (20 cm) wide, pinkish to ochre cream, lilac or purplish-pink at the tips of its shoots.*
- ⊙ **Stem:** *thick and often short, fleshy, whitish to ochre; spore print ochre.*
- ⊙ **Flesh:** *white, thick; sweet flavor and pleasant odor.*
- ⊙ **Habitat:** *on the ground under deciduous and coniferous trees.*
- ⊙ **When to harvest:** *Summer and autumn.*

Genus *Ramaria* is comprised of many species that are often difficult to differentiate from one another, even for experienced mycologists. The clustered coral may be an exception to this rule, but in the UK and North America, where it has been reportedly found, there are concerns that it is in a complex of species, some growing under deciduous trees, others under conifers. Caution is urged. With its robust appearance and its short and dense shoots, delicately tinted a lilac color at their tips, it is fairly easy to recognize. *Ramaria subbotrytis*, re-ported from North America and the UK, is considered inedible without being poisonous, is smaller, and a magnificent coral pink all over.

ear (p. 121); the *Calocera* (p. 279) are part of an entirely separate family and the *Xylaria* (p. 286) are close to the *Peziza*!

Did you know?

Evolution has more than one trick up its sleeve and knows just how to lure mycologists into a trap. *Clavaria* morphology has appeared several times over the course of mushroom history in groups that are very distant from each other. Genus *Ramaria* belongs to the same family as the pig's

Summer
truffle

Tuber aestivum Vittad.

- ◉ **Mushroom in ball shape:** *more or less uniform in shape, subterranean, from 0.75 inches (2 cm) to 4 inches (10 cm) wide, brown-black to black surface sprinkled with pyramidal warts 2–5 mm wide.*
- ◉ **Flesh:** *firm, light brown containing an array of white veins; sweet flavor and faint, somewhat musky odor.*
- ◉ **Habitat:** *under various deciduous trees, particularly oaks, hazels, and hornbeams.*
- ◉ **When to harvest:** *Late spring and summer (May to August).*

other specialists insist upon drawing a clear distinction between the two "species" using a list of differing morphological and organoleptic characteristics (the Burgundy truffle is much more fragrant, among other things). It is difficult to determine whether these debates are grounded in scientific objectivity or passion for these legendary organisms.

Did you know?

Truffles belong to the same large family as the *Peziza*, the morels, and the *Helvella*. Their spores are formed in sacs within their flesh that are located in the brown veins that can be seen clearly when the truffle is cut. These spores are not easily released the way they are in other mushrooms that expose their fertile tissues to the open air. In truffles, spores can only be liberated when the mushroom rots or is unearthed and consumed by animals—like humans— looking for a gourmet treat.

Reported from the UK but not from North America, the summer truffle's existence is controversial: many mycologists think that this name is being used to designate what are actually immature specimens of the Burgundy truffle (*T. uncinatum*), which is harvested in autumn and the beginning of winter. A few genetic tests have served to bolster this hypothesis. Nevertheless,

- ⊙ **Mushroom in ball shape:** usually irregular in shape, humped or lobed, subterranean, from 0.75 inches (2 cm) to 6 inches (15 cm) wide, yellowish ochre surface, sometimes with greenish specks, covered in small bumps.
- ⊙ **Flesh:** firm, whitish at first then reddish-gray with indistinct whitish veins; sweet flavor and a very strong, penetrating odor reminiscent of certain soft cheeses and garlic.
- ⊙ **Habitat:** under various deciduous trees in the Mediterranean.
- ⊙ **When to harvest:** Autumn and early winter.

White
truffle

Tuber magnatum Pico

While this truffle is less well-known than the black truffle, it is nevertheless recognized by gourmet diners as the best truffle in the world. Its intense garlic odor may put many people off, but its flavor is among some of the most delicate. It may be found in markets in North America and the UK but has not been found in the wild in either location.

Warning

There are other subterranean mushrooms that resemble the white truffle but do not possess any of its characteristic qualities. *Tuber borchii*, a synonym of *T. canaliculatum*, reported from northeastern North America, is an exception: it is reported to be a choice edible: it measures up to 2 inches (5 cm) in diameter, its surface is smooth, and its odor, while pleasant, is more fungal. *Choiromyces venosus*, not reported from the UK nor North America, can grow to 4 inches (10 cm) wide, gives off a strong aromatic odor, and its flesh is rather pale and brownish with fairly prominent white veins. Identifying these mushrooms—called hypogeum fungi from the Greek hupo, "beneath, under," and gê, "Earth"—usually requires the use of a microscope.

In the kitchen

The white truffle is a specialty of the Piedmont region in Italy, and as a result many of the recipes dedicated to it are inspired by Italian culture. But be careful: never cook a truffle! They can only be used raw, perhaps grated or in thin slices, incorporated into scrambled eggs, in a risotto, on tagliatelle pasta, or on bread toasted just enough to be crunchy. Another option (also possible with the black truffle, p. 159), is to reserve a small piece of truffle and let it marinate in a bottle of oil, preferably one without a strong flavor, though many people use olive oil. The truffle will transmit its flavor to the oil, which can be used to complement all kinds of dishes.

Did you know?

Not all mushrooms with a somewhat regular tuber shape that grow underground are necessarily truffles. In fact, there are many mushroom groups that count hypogeum (subterranean) mushrooms among their ranks and living conditions beneath the soil are so restrictive that they have all adopted the same "tuberoid" morphology. Experts call the process that has led to these fortuitous similarities "convergent evolution."

A well-known example of these convergences is the "fish" morphology, which, for the same reasons related to a restrictive living environment—water, in this case—has shown up in actual fish like herring, but also in mammals such as dolphins and certain long-gone reptiles like the ichthyosaurs.

The record-breaking truffle

It is an understatement to say that the white truffle is famous, not only for the quality of its flavor but also for its price. In comparison, the black truffle (p. 159)—which goes for about $800 a pound during a bad year—looks terribly cheap. Judge for yourself: in 2007, a white truffle weighing 3.3 pounds (1.5 kg) was auctioned off in London for 225,000 euros ($252,500) to a mushroom lover from Chinese Hong Kong. We will never know whether it arrived in China intact or if it released its spores as it rotted in the hold of some private jet.

- **Mushroom in ball shape:** *typically irregular in shape and more or less humped or lobed, subterranean, from 0.75–4 inches (2–10 cm) wide, black to blackish surface sprinkled with pyramidal warts that have truncated peaks.*
- **Flesh:** *firm, black in maturity with thin white veins; sweet flavor and a very strong, penetrating odor that is aromatic and musky.*
- **Habitat:** *under various deciduous trees, particularly oaks, hornbeams, and hazels.*
- **When to harvest:** *Late autumn and winter.*

Black
truffle

Tuber melanosporum Vittad.

The legendary black truffle, rarely reported from the UK and North America, also known as the Périgord truffle or the black diamond, lives up to its reputation. Its flesh has a consistency at once firm and crunchy, and its powerful odor and unique taste make it a delicacy appreciated by many culinary enthusiasts. It can be confused with the Brumale truffle (*T. brumale*), reported rarely from the UK but not from North America, and mesenteric truffle (*Tuber mesentericum*), not reported from North America nor the UK, both of which are edible but a bit less tasty and grow in the same areas during the same period; the former has thick white veins, the latter a warm brown flesh and a characteristic basal indentation.

In the kitchen

When in season, the black truffle is used in every way imaginable, particularly in French cooking: raw in thin slices on toast, sometimes accompanied by foie gras or fleur de sel, and grated or diced to flavor soups, scrambled eggs, pâtés, terrines, or poultry during the holiday season. We must not forget the famous Périgueux sauce, which features shallots, white wine, and chicken broth in addition to the black truffle. Another traditional use of this truffle consists of placing it in a closed container with eggs; the eggs capture its powerful aroma and can then be used in omelettes, for example.

Did you know?

Part of the legend of the black truffle is undoubtedly connected to the enigma that surrounds how it is harvested. One can certainly use pigs to find them—they are naturally fond of this mushroom—or even train dogs for the same exercise. The most enjoyable way to harvest them, though, is probably to follow the flies. Not just any flies, however: truffle flies (*Suillia tuberiperda*). These tiny brown and furry creatures are rather sluggish and have a keen sense of smell, and they deposit their eggs near the truffle so that their larvae can feed on it. This behavior is a stroke of luck for us: follow the fly, and you will find truffles!

Watch out for fakes

*Foreign truffles, most notably the Asian black truffle (*T. indicum*), have started arriving in markets over the past few years. Even though* T. indicum *is smaller, less fragrant, and presents a few divergent morphological features when compared with the black truffle, an amateur could easily be fooled by a less-than-scrupulous vendor offering the Asian black truffle at black diamond prices. But more important than these small acts of fraud are the potential ecological risks that this new truffle poses. In the wild, Asian black truffles happen to be formidable competitors for the indigenous European truffles. Better equipped than its peers to settle on the roots of host trees, it has already been sighted on truffle farms, and many in Europe are wondering if one day it will simply kill off their precious native truffles.*

▼ Fly agaric

Inedible
or
Poisonous
Mushrooms

Prince

Agaricus augustus Fr.

- ⊙ **Cap:** *up to 10 inches (25 cm) wide, covered with tawny, russet, or ochre-brown scales.*
- ⊙ **Gills:** *free and crowded, cream, then gray-pink and finally brown; spore print dark brown.*
- ⊙ **Stem:** *whitish, yellows slightly when bruised, with a large skirt-shaped ring with flakes the same color as the cap.*
- ⊙ **Flesh:** *firm; sweet flavor and pleasant bitter almond odor.*
- ⊙ **Habitat:** *in and around deciduous woods, less frequently beneath conifers.*
- ⊙ **When to observe:** *Autumn.*

This beautiful agaric, reported from the UK and western North America, is not common but is easy to recognize thanks to its majestic stature, its cap sprinkled with colorful scales, its large skirt-shaped ring with an inferior face adorned with enormous triangular flakes around the edge—giving it the appearance of a serrated wheel—and finally by its agreeable bitter almond scent. It is unlikely to be confused with other mushrooms apart from large agarics with smooth caps or scales the same color as the cap. It is very widely consumed in many European countries as well as in the United States, but it has recently been found to contain substances that are most likely carcinogenic. It is probably best, therefore, to either use restraint in the amount consumed or to leave it undisturbed and admire it in its natural habitat.

Warning

Be careful not to confuse the prince with the blushing wood mushroom (p. 48). The latter is a good edible that also has a scaly cap, but it is smaller, browner (less ochre-yellow), and its flesh reddens significantly when cut.

Pavement
mushroom

Agaricus bitorquis (Quél.) Sacc.

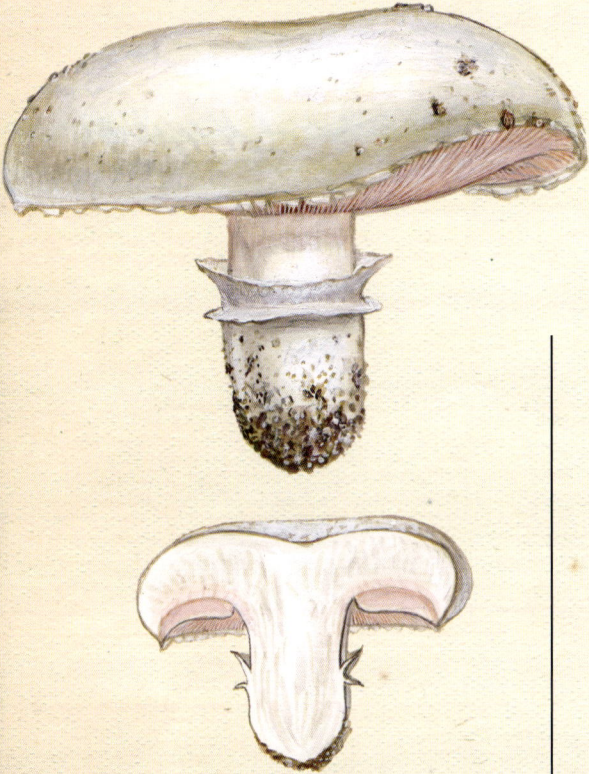

- ◉ **Cap:** *up to 4 inches (10 cm) wide, smooth or somewhat cracked by dry weather, whitish, very firm or even hard in the center.*
- ◉ **Gills:** *free and crowded, pink then brown; spore print dark brown.*
- ◉ **Stem:** *whitish, very firm, with a characteristic double ring.*
- ◉ **Flesh:** *firm, reddening slightly; sweet flavor and mushroom odor.*
- ◉ **Habitat:** *in areas with significant foot traffic like roadsides and hard packed trails; it sometimes even pushes through sidewalks!*
- ◉ **When to observe:** *Spring to autumn.*

Reported from North America and the UK, the pavement mushroom has a somewhat controversial reputation. It is considered an edible mushroom in several countries, but it has an annoying tendency to grow in polluted areas rich in pesticides, nitrate fertilizers, and other heavy metals that it greedily devours. These compounds, which are hardly beneficial to our health, accumulate in the pavement mushroom's tissues and it would be unwise to consume it. That being said, specialists have known how to cultivate it since the late 1960s, so it is not entirely impossible to imagine finding pollution-free strains on our dinner plates someday. So, except for possible environmental toxins, it is considered edible.

Warning

Be careful not to confuse the pavement mushroom with the meadow mushroom (p. 45), which grows in grass and has softer flesh, a stem that is pointed at its base, and a thin and delicate ring that often disappears in adult specimens. *Agaricus bresadolanus* (p. 164) grows in the same areas, as well. *Agaricus bernardii* is similar, but it grows in sandy soil, its flesh stains red, and smells salty. Considered edible, it is reported from coastal areas of North America and the UK.

Agaricus
bresadolanus

Bohus

- ◉ **Cap:** *up to 4 inches (10 cm) wide, covered in brownish fibrillose patches in a radiating pattern, sometimes rather sparse, on a whitish background.*
- ◉ **Gills:** *free and crowded, gray-pink then brown; spore print brown.*
- ◉ **Stem:** *whitish, club-shaped and typically elongated by a small hard "root" at the base, yellowing slightly when bruised, with a whitish ring that is not very thick.*
- ◉ **Flesh:** *yellowing somewhat at the base of the stem; sweet flavor and faint almond odor.*
- ◉ **Habitat:** *in and around deciduous woods, along roadsides and trails, sometimes also in meadows.*
- ◉ **When to observe:** *Spring to autumn.*

This small and rather common agaric, reported from the UK but not from North America, is often confused with the meadow mushroom (p. 45) because of its white and almost bare cap. It is fairly difficult to digest because it often grows in polluted soils. To prevent making a mistake in identification, note that *Agaricus bresadolanus* has a club-shaped stem with a small root—while the stem of the meadow mushroom is fusoid without a root—and remember that its ring is well-formed, not fine and evanescent like its edible cousin's.

Did you know?

A few years ago, mycologists separated *A. bresadolanus* (which supposedly grew only in meadows) from *A. romagnesi*, which preferred areas rich in nitrogen compounds and often chose spots along major roads and trails. Recent and meticulous studies using molecular biology techniques have proven that only one species exists, *A. bresadolanus;* it just happens to be a species that is morphologically variable and fairly undemanding when it comes to its habitat.

Inky
mushroom

Agaricus moelleri Wasser

- ⊙ **Cap:** *up to 6 inches (15 cm) wide, completely covered in delicate gray or charcoal gray scales that spread apart from each other as the cap grows, allowing a glimpse of the whitish background that sometimes turns yellow when bruised.*
- ⊙ **Gills:** *free and crowded, gray-pink then brown; dark brown spore print.*
- ⊙ **Stem:** *whitish, fairly long, often bulbous, yellows significantly at the base, white ring that is not very thick.*
- ⊙ **Flesh:** *yellows significantly at the base of the stem; sweet flavor and iodine or ink odor.*
- ⊙ **Habitat:** *in deciduous forests.*
- ⊙ **When to observe:** *Especially in autumn.*

The inky mushroom, reported from the UK but rarely from North America, is part of the yellow stainer mushroom group (p. 167) and is not edible. It usually grows in dark undergrowth with soil rich in humus and has a high tolerance for pollution, which explains why it is often found in urban forests and parks. The small gray spots on its cap and its yellowing flesh are its characteristic features, as is the iodine odor it shares with all agarics in its group, an indicator that these species are inedible.

Another species

Today genus *Agaricus* includes around 100 species in Europe, and many of them are difficult to tell apart. *A. parvitigrinus*, one of the yellow stainers with an iodine odor—a group mycologists refer to as *Xanthodermatei*—is a miniature version of the inky mushroom whose cap does not grow more than 2.5 inches (6 cm) wide. It was described for the first time in 2005 during harvests in the Gironde region of France; there are no reports of this species from North America or the UK.

- **Cap:** up to 6 inches (15 cm) wide, often has an irregular shape, truncated then flat, white (see box on next page for variations), yellowing when bruised.
- **Gills:** free and crowded, bright pink then brown; spore print dark brown.
- **Stem:** white, often distinctly bulbous and yellowing, fine ring.
- **Flesh:** white, turns yellow; unpleasant flavor and sweetish iodine odor.
- **Habitat:** on the edge of forests, in parks and gardens.
- **When to observe:** All year long, apart from periods of extreme cold, but especially in autumn.

Yellow
stainer

Agaricus xanthodermus (Scop. : Fr.) Pers.

Variations on the yellow stainer...

Mushrooms can be quite perplexing for an amateur because they sometimes appear in such a variety of forms that it can feel impossible to narrow down the possibilities to a single species. The yellow stainer, reported from the UK and North America, is no exception to this rule, and while it typically grows with a white cap that some may describe as "ordinary," it can also be found with a distinctly gray cap (var. griseus, p. 168) or covered with small brown scales that cause it to resemble certain Lepiota (var. meleagroides). Ascertaining the variations within mushroom species and being able to recognize them is one of the greatest difficulties in the field of mycology.

The yellow stainer is one of the most common agarics. Even though some people are able to consume it without any problem, it is important to know how to recognize it because in most cases it causes what is known as resinoid syndrome. Symptoms include nausea, vomiting, and abdominal pain that can persist for long periods of time in highly sensitive individuals. Fortunately, this mushroom can be readily identified by the way its flesh turns yellow: just break off a small piece of the bulb to see if the flesh rapidly changes to a bright yellow and check for a distinct and somewhat unpleasant iodine odor. It should be noted that this odor becomes accentuated during cooking, so if this mushroom happens to have landed in the pan accidentally, its characteristic smell is an excellent indication that an incorrect identification may have taken place.

Warning

The yellow stainer is most likely to be confused with the wood mushroom (p. 49), which does turn yellow but less intensely and in a slower manner. The latter also gives off a pleasant aniseed odor, which is a significant difference. When the yellow stainer grows in meadows and pastures, it may be mixed in with the meadow mushroom (p. 45), and you will need to remember that the latter does not turn yellow, its stem is not bulbous and bears an extremely fragile ring, and it exudes an agreeable mushroom scent.

Other species

Yellow stainers with an iodine smell are grouped by mycologists in section *Xanthodermatei*. The species within this group—a list that is currently under revision by specialists—are large in number and often difficult to tell apart. *A. menieri* for instance is a yellow stainer looka-

▲ *Agaricus menieri*

like that grows in the sand dunes of the European Atlantic coast and Mediterranean. *A. xanthodermulus*, not reported from North America or the UK, is a small yellow stainer that grows in sandy soil.

Yellow stainer
(var. griseus)

Agaricus xanthodermus var. griseus (A. Pearson) Bon & Cappelli

- ⊙ **Cap:** up to 6 inches (15 cm) wide, often irregular in shape, truncated then flattened out, uniformly gray in most cases, yellows when bruised.
- ⊙ **Gills:** free and crowded, bright pink then brown; spore print dark brown.
- ⊙ **Stem:** white, often distinctly bulbous, turns unmistakably yellow, thin ring.
- ⊙ **Flesh:** white, yellowing; unpleasant flavor and sweetish iodine odor.
- ⊙ **Habitat:** on the edge of forests, in parks and gardens.
- ⊙ **When to observe:** Most of the year, apart from periods of extreme cold, but especially in autumn.

This agaric was once considered its own species but is actually just a gray-capped variety of the yellow stainer (p. 167), reported only from France. It is therefore extremely difficult to digest and should not be consumed. Its unpleasant and even repellent odor, among other traits, allows most people to recognize it without too much difficulty. Its flesh turns intensely yellow, particularly in the base of the stem.

Did you know?

This agaric was described for the first time in 1946 by English mycologist Arthur Anselme Pearson as *Psalliota xanthoderma* var. *grisea*.

The genus name *Psalliota*, which was once used for all agarics, was abandoned for obscure reasons in favor of genus *Agaricus*, and this variety's current name was bestowed upon it in 1983 by two other mycologists, Marcel Bon and Alberto Cappelli. This change of name is summarized after the Latin name (see above) as "(A. Pearson) Bon & Cappelli."

False
death cap

Amanita citrina (Schaeff.) Pers.

- ◉ **Cap:** *reaches 4 inches (10 cm) in diameter, pale lemon yellow with irregular patches the same color or a little browner.*
- ◉ **Gills:** *free and crowded, pale lemon yellow like the cap; spore print white.*
- ◉ **Stem:** *has a thin membranous ring and a large abruptly bulbous bulb, all of which is pale lemon yellow like the cap.*
- ◉ **Flesh:** *yellowish; distinct flavor and odor, resembling those of radishes and raw potatoes.*
- ◉ **Habitat:** *under deciduous and coniferous trees.*
- ◉ **When to observe:** *Most of the year, apart from periods of extreme cold, but especially in autumn.*

Here we have one of the most common *Amanita*. Reported from eastern North America and the UK, there is some debate among mycologists concerning con-specificity of the European and North American populations, but it is one of the easiest to recognize: it is colored all over in its typical shade of pale yellow, its stem bears a large marginate bulb with a hemisphere shape, and its flesh gives off a strong radish and raw potato odor. The false death cap could potentially be confused with the jonquil amanita (p. 170), which also has a yellow cap, but the gills and stem of the jonquil amanita are white, and its flesh releases a faint odor that is hardly distinctive. The European star-footed amanita is distinguished by its cap, which quickly stains a rust color when bruised or attacked by insects, and by its bulb, which is well-formed but less marginate than that of the false death cap and tends to split lengthwise in a star shape.

Did you know?

The false death cap is sometimes all white. This deceiving form, which often grows blended in among "normal" specimens, nevertheless possesses all of the other features typical of this species; it is known as the *alba* form, from the Latin word *alba*, meaning "white."

▲ *Amanita citrina* f. *alba*

Jonquil
amanita
Amanita junquillea Quél.

- **Cap:** up to 4 inches (10 cm) wide, ochre-yellow to orangey-yellow, often with irregular patches from the white veil.
- **Gills:** free and crowded, white; spore print white.
- **Stem:** white with a thin ring that is membranous, fragile, and often disappears in adult specimens; a rounded bulb with a few ridges.
- **Flesh:** white; sweet flavor and faint odor.
- **Habitat:** especially under conifers, rarer under deciduous trees.
- **When to observe:** Especially in autumn.

This elegant *Amanita*, reported from North America and the UK, is easy to recognize because its rather dashing yellow contrasts with its white gills and stem. Its bulbous stem bears a ring that is most visible in young and fresh specimens. The ring is extremely fragile and tends to disappear when the mushroom grows larger and ages. It may be confused with the false death cap (p. 169), which can be distinguished by its gills and stem that are the same color as the cap and by the strong radish odor of its flesh.

Did you know?

The history of species names is often quite complex, and untangling its mazes is the task of an entirely separate branch of natural science: nomenclature. The jonquil amanita has received many other names over the years, including *Amanita sulphurea*, *Amanita lutea*, *Amanita adnata*, and even *Hypophyllum nitido-guttatum*, and may now be called *Amanita gemmata*, and possibly *Amanita russuloides* in North America. Specialists sometimes battle for years to decide which name to adopt from the many that abound in scientific literature.

Amanita proxima

Dumée

- **Cap:** *reaches 8 inches (20 cm) in diameter, smooth, white or cream to pale beige.*
- **Gills:** *free, crowded, white or cream; spore print white.*
- **Stem:** *white like the gills, emerging from a large membranous volva that is typically a russet color, has a wide white or reddish ring that is fairly membranous.*
- **Flesh:** *white; sweet flavor and a "from the sea" iodine odor that becomes unpleasant as the mushroom ages.*
- **Habitat:** *especially under oak trees and in calcareous soil in Southern Europe.*
- **When to observe:** *Late summer and in autumn.*

Easily confused with the bearded amanita (p. 52), *Amanita proxima*, not reported from the UK or North America, is toxic, so it is important to be able to recognize it. This mushroom's volva is tinted a distinct russet color. The bearded amanita's volva, on the other hand, is white or only somewhat russet-colored. The *Amanita proxima* stem also bears a well-formed ring that is fairly membranous while still fragile; in its edible cousin, the ring is almost indistinct and has a consistency like whipped cream. *A. proxima* is also smaller and less fleshy than the bearded amanita, but this distinction is not always reliable and very large specimens are sometimes found.

Other species

There are plenty of other large white *Amanita* species and distinguishing them sometimes requires a little bit of experience. Among the most common we might mention are the warted amanita (*A. strobiliformis*), reported from the UK and eastern North America, not edible, which has a cap covered in large irregular patches and a creamy ring similar to the one seen in the solitary amanita (*A. echinocephala*), reported from the UK and rarely from eastern North America, which is also inedible and has a membranous ring as well as a cap and stem base adorned with truncated pyramidal warts.

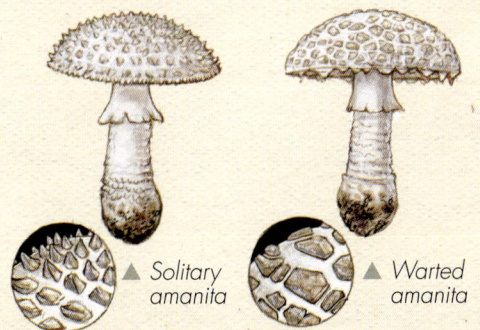

▲ Solitary amanita

▲ Warted amanita

- ◉ **Cap:** *reaching 8 inches (20 cm) in diameter, typically a beautiful red or orangey-red and covered with white flakes that may disappear as the mushroom ages or is subjected to bad weather.*
- ◉ **Gills:** *free and crowded, white; spore print white.*
- ◉ **Stem:** *white, often distinctly bulbous, with a white ring and a bulb topped with protrusions or white scales on the bulb.*
- ◉ **Flesh:** *white; faint flavor and odor.*
- ◉ **Habitat:** *under deciduous and coniferous trees.*
- ◉ **When to observe:** *Summer but especially in autumn.*

Fly agaric

Amanita muscaria (L. : Fr.) Lamarck

Astonishing properties...

The fly agaric was widely used by Siberian shamans for its hallucinogenic properties. The rituals can hardly be described as appetizing: the shaman would typically consume the fly agaric and the people being initiated would then drink his urine. Unlike the toxic molecules that cause excessive perspiration and violent gastrointestinal reactions, which were "filtered out" by the shaman's body, psychoactive substances are concentrated in urine. It should also be noted that fly agaric toxins are water-soluble, and certain field guides that we might qualify as "extremist" encourage the consumption of this mushroom after cooking it for a long period at high heat in plenty of water. One wonders what could possibly justify taking such a risk . . .

The fly agaric is legendary and is perhaps the most well-known mushroom among children. Strangely, it is often this mushroom that people describe when asked to talk about the death cap (p. 229), a very different species and far more toxic. The fly agaric grows in very distinct habitats, does not like calcareous soil, and adjusts to the company of many different trees while showing a preference for birches, pine, and spruce. It is present all over the world and usually follows one of its primary hosts, the pine tree, into every country where it is introduced.

Warning

Commonly reported in North America and the UK, the fly agaric is difficult to confuse with other mushrooms, and its white gills and stem, its cap covered with white warts, and its lack of a membranous volva are distinctive features. Many mushroom lovers, though, are wary of its resemblance—however superficial—to the Caesar's mushroom (p. 51). Just remember that the Caesar's mushroom emerges from a large sac volva that persists at the base of the stem, and that its gills, stem, and ring are a lovely yellow color.

Other species

In Europe, the fly agaric may display a variety of very different and sometimes confusing forms. The *aureola* variety has a bright orange cap with hardly a flake at all. The *flavivolvata* form has a bright yellow veil, and as a result has flakes and protrusions tinted the same color, while in the *fuligineoverrucosa* variety they are a sooty gray. There are even some specimens—the *alba* variety—that are completely white. In northeast North America the cap may be more yellow (var. *guessowii*), but var. *alba* is also reported. In the UK if the cap is yellow it may be var. *formosa*.

▲ *A. muscaria*
var. aureola

▲ *A. muscaria*
var. flavivolvata

▲ *A. muscaria*
var. alba

- ◉ **Cap:** *reaching 4 inches (10 cm) in diameter, striated edge, brown to brownish-beige, covered with irregular white flakes.*
- ◉ **Gills:** *free and crowded, white; spore print white.*
- ◉ **Stem:** *white with a distinct bulb that has a "turtleneck" border, a few protrusions and a membranous ring that is not striated on the superior face.*
- ◉ **Flesh:** *white; faint flavor and odor.*
- ◉ **Habitat:** *beneath deciduous and coniferous trees.*
- ◉ **When to observe:** *Summer but especially in autumn.*

Panther
cap

Amanita pantherina (DC. : Fr.) Krombh

A little bit of history

It was the great Swiss botanist Augustin Pyrame de Candolle who first described the panther cap as Agaricus pantherinus while helping his colleague Jean-Baptiste de Lamarck reshape his Flore française ("French Flora") in 1815. It was not until 1843 that German mycologist Julius Vincenz von Krombholz assigned this mushroom to genus Amanita in his renowned work entitled Naturgetreue Abbildungen und Beschreibungen der essbaren, schädlichen und verdächtigen Schwämme, which can be translated as "Faithful illustrations and descriptions of edible, noxious, and suspect mushrooms."

This mushroom is a beautiful and elegant *Amanita* that is highly toxic but far more attractive than its edible relative, the blusher (p. 53). The panther cap can be recognized by its cap, which is striated along the edge, relatively dark brown, and flecked with bright white flakes. Its stem is also distinctive, with a characteristic bulb shape and a rim rolled over on itself like a turtleneck, often decorated with a few white wreaths. The gills of the panther cap are white, and its white flesh is almost completely odorless and does not redden. It is common and can be found in both deciduous and coniferous forests. The panther cap's *abietum* form, which is darker and has fewer striations around the edge of the cap, can sometimes be found under coniferous trees.

Warning

While the panther cap is reported from across North America and the UK, and easy to recognize if you account for each element described above, it is still sometimes confused with the blusher (p. 53), which can be distinguished from the panther cap by the reddening flesh that gives it its characteristic "dirty" appearance, its irregular flakes that are more gray than white, and by its stem, which terminates in an onion-shaped bulb without a well-defined edge. In North America, both *Amanita multisquamosa* and *velatipes* may be confused with the panther cap. The gray spotted amanita (*Amanita excelsa* var. *spissa*), reported from the UK but not North America, is also inedible and resembles the panther cap, but its irregular flakes are gray, its bulb has an onion shape, and its flesh gives off a turnip scent.

Did you know?

Limacella species are very closely related to the *Amanita* but can be differentiated by their universal veil (later giving way to the volva), which is not membranous or flaky but instead viscous and sticky. This feature gives these mushrooms—which release an odor that is often distinctly flour-like—a singular appearance. The weeping slimecap (*Limacella guttata*), a good edible in some regions (refer to the discussion in the edible section of this book), is known not only for its viscous cap, but also its pale ochre-pinkish-beige hues and its lovely membranous ring. More abundant in mountainous regions, it grows under deciduous and coniferous trees.

Chlorophyllum
brunneum

(Farl. & Burt) Vellinga

- ◉ **Cap:** *reaches 6 inches (15 cm) in diameter, covered in brown to brownish-gray scales, with a well-defined star-shaped center on a white or whitish background.*
- ◉ **Gills:** *free, white to cream; spore print white.*
- ◉ **Stem:** *smooth, reddening when scratched, usually with a large marginate bulb and a thick sliding ring that is plain in appearance.*
- ◉ **Flesh:** *whitish, turning bright orangey-red when cut; faint flavor and odor.*
- ◉ **Habitat:** *in bushes, parks, and gardens, often in soil enriched with nitrates.*
- ◉ **When to observe:** *Summer and autumn.*

This large *Lepiota*, reported from North America and the UK, is not very common and often grows in rich soil found in gardens or compost piles. Its reddening flesh is reminiscent of the shaggy parasol (p. 88), but it is known for having a paler cap that stands out beneath its brown scales and a stem with a voluminous marginate bulb. It has caused a few mild reactions with symptoms that resemble a stomach virus (abdominal pain, nausea, vomiting, etc.), and it appears to be more indigestible than truly toxic, though some people seem to have an allergic reaction.

Did you know?

Up until a few years ago, all of the large European *Lepiota* were classified under genus *Macrolepiota*, and genus *Chlorophyllum* was reserved for a few toxic tropical species, many of which have a green spore print. Recent studies have shown, however, that *Chlorophyllum brunneum*, like the shaggy parasol (*Chlorophyllum rhacodes*), is closer to these tropical *Chlorophyllum* species than to other large *Lepiota*. Its genus has been changed out of respect for these evolutionary connections.

Frosty
funnel

Clitocybe phyllophila.(Pers. : Fr.) P. Kumm.

- **Cap:** *reaches 4 inches (10 cm) in diameter, fairly elastic, whitish with a bloom that disappears when rubbed with a fingertip, allowing the ochre background to appear.*
- **Gills:** *adnate or slightly decurrent, fairly crowded, cream to pale ochre-beige; spore print cream to pinkish.*
- **Stem:** *fairly short, the same color as the cap.*
- **Flesh:** *whitish, fairly elastic; sweet flavor and faint odor, somewhat flour-like or earthy.*
- **Habitat:** *in hedges and bushes, often on decomposing leaves.*
- **When to observe:** *Especially in autumn.*

Out of all of the toxic *Clitocybe* species, the frosty funnel (reported from across North America and the UK) is perhaps the easiest one to confuse with the sweetbread mushroom (p. 62): the two mushrooms have the same silhouette and grow in the same areas. To avoid making a mistake, it is very important to note that the sweetbread has very fragile and brittle flesh that releases a strong odor of fresh flour. Its distinctly decurrent gills are pink when mature (pink spore print). The frosty funnel has tougher, more elastic flesh, a faint odor, and gills that are not decurrent and usually a cream or pale ochre color in mature specimens.

Did you know?

Around 300 *Clitocybe* species exist worldwide, and many of them are difficult to identify, even for experienced mycologists. Even their microscopic features are fairly uniform, and identification hinges on subtle differences in the shape of spores and other cells that make up the mushrooms' flesh. The color of the spores—evaluated using a spore print, of course—is also very important. There are several other genera that are very similar to these mushrooms, which only complicates the mycologist's task further: genus *Lepista*, for example, includes species whose spores are often festooned with tiny warts, while those of the *Clitocybe* are smooth.

- **Cap:** *up to 3 inches (8 cm) wide, rather fleshy, ochre to reddish-beige or pinkish-ochre.*
- **Gills:** *decurrent, cream then reddish-beige.*
- **Stem:** *whitish to pinkish-ochre.*
- **Flesh:** *white; sweet flavor and strong odor, reminiscent of orange blossom flower.*
- **Habitat:** *beneath conifers, usually in calcareous soil.*
- **When to observe:** *Autumn.*

Paralysis
funnel

Clitocybe (Paralepistopsis) amoenolens Malençon

You've got to have the nose for it!

Mycology is a difficult field because mushrooms are morphologically quite variable and possess only a few characteristics to cling onto when trying to recognize and correctly identify them. Odor and flavor are features that are frequently used and distinctive in many mushrooms, but they can still prove quite challenging for a beginner. While flavor is relatively easy to describe (sweet, bitter, tangy, etc.), describing an odor is often subjective, and becoming a mycologist means undergoing training . . . of one's nose. With a little experience, you will be able to detect characteristic odors like fresh flour, cucumber, mandarin, water, oyster, a decaying corpse . . . even coconut cake and engine smoke!

The paralysis funnel is not reported from the UK or North America. Still, it is important to know how to recognize it because in the mid-1990s it was responsible for serious and very painful cases of acromelalgia syndrome. A few days after ingesting the mushroom, poisoned individuals watch as their extremities become red and swollen. This transformation is accompanied by unbearable burning sensations that are difficult to relieve even with analgesics and frequent ice baths. The toxic molecules (acromelic acids) are powerful neurotoxins and their effects can last for several weeks.

Warning

If you examine it superficially, the paralysis funnel could easily be confused with the tawny funnel mushroom, a rather uninteresting species that is nevertheless consumed in certain regions. The tawny funnel has a less fleshy cap that is sunken in a funnel shape in the center and flesh that is almost odorless. Also avoid confusion with edible *Clitocybe* species from the monk's head group (p. 59), which release a faint odor of cut grass.

Did you know?

The paralysis funnel was first described as *Clitocybe fallaciosa* by well-known mycologist Georges Malençon during a harvest in Morocco's Azrou-Ifrane region in 1959. Unfortunately, this name had already been used by another mycologist for different species, forcing Malençon to replace his first choice with *Clitocybe amoenolens* in his monumental book *Flore des champignons supérieurs du Maroc* ("Flora of Superior Mushrooms of Morocco"), which he published in 1970 in collaboration with another French mycologist, Raymond Bertault.

- ◉ **Cap:** reaching 2.5 inches (6 cm) in diameter, fairly elastic, whitish, with a bloom that disappears when rubbed with a fingertip, allowing the ochre background to appear.
- ◉ **Gills:** adnate or slightly decurrent, fairly crowded, cream to pale ochre-beige; spore print white.
- ◉ **Stem:** fairly short, the same color as the cap.
- ◉ **Flesh:** whitish, elastic, sweet flavor and faint odor, sometimes a bit like flour.
- ◉ **Habitat:** in the grass of lawns and prairies.
- ◉ **When to observe:** Summer and autumn.

Fool's
funnel

Clitocybe rivulosa (Pers. : Fr.) P. Kumm.

The Clitocybe that causes weight loss...

Don't get too excited; you are not going to find a miracle recipe for losing a few extra pounds in this section, but you will find plenty of information about the symptoms of toxic Clitocybe poisoning. Consuming these mushrooms unfortunately causes generalized hypersecretion, a massive release of tears, sweat, saliva, etc., that is made even more uncomfortable—if that wasn't bad enough—by a significant drop in blood pressure, diarrhea, nausea, and vomiting. It's certainly one way to spend a few unforgettable hours and lose a few pounds of water.

This small *Clitocybe*, reported from the UK and North America, often grows abundantly on lawns and prairies where it sometimes forms large fairy rings. Its cap's appearance is difficult to describe but is typical of all of the toxic *Clitocybe* species and is sometimes said to be "frosted" or "iced." It does, in fact, appear to be covered in a fine layer of frost or powdered sugar that lets the ochre-brown background peek through in patches or regular streaks as it disappears. Its fairly crowded gills are pale with a slight ochre tint that is typical, and its flesh gives off a faint odor that is not distinctive.

Warning

Lovers of the Scotch bonnet mushroom (p. 90) should be especially attentive when searching for their favorite species: the fool's funnel has an irritating tendency to form fairy rings in the same areas as the Scotch bonnet, and highly toxic *Clitocybe* sometimes grow right in the midst of these delicious edible mushrooms. It is therefore imperative to thoroughly examine each specimen, keeping in mind that the Scotch bonnet has spaced gills that are distinctly emarginate, a remarkably tough stem, and is more ochre in color.

Did you know?

The *Clitocybe* belong to the large *Tricholomataceae* family. This very diverse group includes saprophytic mushrooms like the *Clitocybe*, mycorrhizal mushrooms like the *Tricholoma*, and even true lichens such as the *Lichenomphalia* (thanks to their symbiotic relationship with algae, these small mushrooms that resemble small *Clitocybe* with spaced gills grow on substrates that are very poor in nutrients such as sphagnum moss and in highly acidic soil. These lichens are most plentiful in alpine and arctic zones but are sometimes found on bare ground in the lowlands).

Common
inkcap

Coprinopsis atramentaria (Bull. : Fr.) Redhead et al.

- ⦿ **Cap:** *2.75 inches (7 cm) wide, conical with an obtuse peak, silvery gray, smooth or delicately scaly.*
- ⦿ **Gills:** *ascending and free, white then grayish and finally black, deliquescent.*
- ⦿ **Stem:** *white with a small ring-shaped zone with protrusions at its base; spore print black.*
- ⦿ **Flesh:** *white, very deliquescent; sweet flavor and faint odor.*
- ⦿ **Habitat:** *often in tufts on lawns and along trails.*
- ⦿ **When to observe:** *Spring to autumn.*

This *Coprinopsis* species is very common, reported from North America and the UK, and easily recognizable because of its considerable size and its conical cap that is blunt on top, which sets it apart from close species like the humpback inkcap (*C. acuminatus*), reported from the UK but rarely from North America, or *Coprinopsis romagnesiana*, reported from North America and the UK, both of which have caps with pointier peaks. The humpback inkcap is also smaller, and *C. romagnesiana* has a cap covered with small ochre-brown to russet brown scales. None of these *Coprinopsis* species are edible, and care should be taken not to confuse them with the shaggy inkcap (p. 69), whose cap has the shape of a gloved finger covered with small wooly white scales.

▲ Humpback inkcap   ▲ *Coprinopsis romagnesiana*

Did you know?

Astonishingly, the common inkcap is only toxic when its consumption is accompanied by alcohol. When this happens, a series of symptoms follows. While these symptoms are generally not life-threatening, they are extremely uncomfortable: reddening of the face, heartbeat irregularity, migraines, dizzy spells, etc. These problems are similar to those experienced by patients undergoing treatment for chronic alcoholism (known as the "Antabuse" effect) and may reappear over several days if the individual consumes more alcohol.

Magpie
inkcap

Coprinopsis picacea (Bull. : Fr.) Redhead et al.

- ◉ **Cap:** conical egg shape, covered in white, grayish, or pink cottony patches that break apart allowing the black background to appear.
- ◉ **Gills:** ascending and free, white then grayish and finally black, deliquescent.
- ◉ **Stem:** white, flaky.
- ◉ **Flesh:** white, very deliquescent; sweet flavor and distinct asphalt odor.
- ◉ **Habitat:** in deciduous forests, often under beech trees.
- ◉ **When to observe:** Spring to autumn.

The magpie inkcap, reported from the UK and North America, is very easy to recognize and cannot be confused with any other mushroom. Its black cap adorned with white blotches resembling a magpie's plumage—the reason for its name—and its highly deliquescent flesh with an asphalt odor are all unique features. It grows most often in the woods in small, isolated groups. The shaggy inkcap (p. 69) resembles it a little as it ages and slowly begins to liquefy, but its cap is covered with wooly white scales, its stem has a white ring at its base, and it does not have an odor; it also usually grows in grass.

Did you know?

The magpie inkcap was first described by Jean-Baptiste François Pierre Bulliard in 1785 as *Agaricus picaceus*. At the time, the majority of mushrooms with gills were dumped into genus *Agaricus*, which today only contains species related to common button mushroom (p. 46). It was placed in genus *Coprinus* in 1821 and moved again to genus *Coprinopsis* in 2001. These kinds of name changes in the vast and already complicated world of mycology present yet another challenge for amateurs who have a hard time grasping the often-subtle reasoning behind these upheavals.

Violet
webcap

Cortinarius violaceus (Fr. : Fr.) S. F. Gray

- ⊙ **Cap:** *reaches 6 inches (15 cm) in diameter, convex, a beautiful dark violet-blue and typically velvety.*
- ⊙ **Gills:** *emarginate, dark violet-blue, rust-colored speckling from the spores in adult specimens; spore print rusty brown.*
- ⊙ **Stem:** *often club-shaped, the same color as the cap, with a cortina that is rust-colored from the spores.*
- ⊙ **Flesh:** *dark purple; sweet flavor and distinct odor called "Russian leather."*
- ⊙ **Habitat:** *under deciduous and coniferous trees.*
- ⊙ **When to observe:** *Late summer through autumn.*

The violet webcap, reported from North America and the UK, is easy to recognize, which is not always the case among *Cortinarius* species. With its full violet blue coloring, its remarkably felty cap, and its characteristic odor—one that is nevertheless difficult to describe, though some mycologists compare it to thyme—it can hardly be mistaken for any other mushroom. There is, however, one lookalike that grows under spruces and can only be distinguished (if not by its habitat) by the size of its spores. Needless to say, identifying it is usually a task left to the experts.

Did you know?

Cortinarius are legendary among mycologists! The genus is so vast (there are probably over 1,000 species in North America) and the species that make it up are often so difficult to tell apart that very few specialists are brave enough to study them. The most difficult ones to recognize are the small brown *Cortinarius* species.

The extent of the diversity within genus *Cortinarius* is still largely unknown, even in Europe, and these mushrooms will continue to be the object of scientific study in years to come.

Wood
pinkgill

Entoloma rhodopolium (Fr. : Fr.) P. Kumm.

- ⊙ **Cap:** *reaches 4 inches (10 cm) in diameter, not fleshy and often wavy, grayish-brown, becomes paler as it dries.*
- ⊙ **Gills:** *emarginate, whitish then pinkish; spore print pinkish-brown.*
- ⊙ **Stem:** *fairly slender and somewhat fragile, whitish or grayish.*
- ⊙ **Flesh:** *whitish; sweet flavor and faint odor, sometimes a sharp bleach odor.*
- ⊙ **Habitat:** *beneath deciduous trees.*
- ⊙ **When to observe:** *Summer and autumn.*

The wood pinkgill is very common, reported from the UK and North America, and only grows in forests, which should make it easier to identify. Its pale and hygrophanous cap (meaning it changes color as it dries, usually fading) and its typically odorless flesh complete its portrait. Once in a while it gives off a distinct and unpleasant odor, one that is a bit sharp and similar to bleach, and in this case is said to be in its *nidorosum* form, reported from the same geographic range as the species. Several other forest *Entoloma* exist that resemble the wood pinkgill, and telling them apart requires experience and, sometimes, a good microscope. *E. sericatum*, for example, usually grows in humid places under willow trees, and its flesh often exudes a faint flour-like odor. It is reported from the same geographic range as the species.

Did you know?

Entoloma have a pink spore print, but this characteristic alone is not enough to distinguish them from other gilled mushrooms because *Volvariella*, *Pluteus*, and a few other genera also have pink spore prints. Even though *Volvariella* and *Pluteus* species have free gills that no *Entoloma* species possess, differentiating other genera of "rhodospore" mushrooms (from the Greek *rhodon*, "pink") is sometimes a delicate matter requiring a look at their microscopic features. The sweetbread mushroom (p. 62), for instance, also has pink spores, but can be readily identified by the spores' elongated and fluted shape, whereas *Entoloma* spores are distinctly polygon-shaped.

- **Cap:** *up to 6–8 inches (15–20 cm) wide, convex, often with a large dome-shaped umbo, pale gray to yellowish beige, does not significantly change color as it dehydrates.*
- **Gills:** *emarginate, not very crowded, yellow at first then pinkish; spore print pinkish.*
- **Stem:** *fleshy, often club-shaped, white then cream.*
- **Flesh:** *white; distinctly flour-like flavor and odor.*
- **Habitat:** *under deciduous trees, especially in calcareous soil.*
- **When to observe:** *Summer and autumn.*

Livid
pinkgill

Entoloma sinuatum (Bull. : Fr.) P. Kumm.

The history of mycology is riddled with pitfalls...

In their publication entitled Traité des champignons comestibles, suspects et vénéneux qui croissent dans le bassin sous-pyrénéen *("Treatise on edible, suspect, and poisonous mushrooms growing in the Sub-Pyrenean basin"), published in 1838, Drs. J.-B. and A. Dassier describe the livid pinkgill in detail using the common name "Agaric phonosperme." Their description is remarkably precise and accompanied—rightly so, to avoid errors—by a beautifully colored panel of illustrations. They state that the great mycologist J.-B. Bulliard may have confused several close species under this name, which is possible. None of this would be terribly problematic if the two authors had not concluded the section dedicated to this mushroom with the following recommendations: "The Agaric phonosperme is edible and perfectly suited to use in a white sauce or fried in a pan with olive oil or lard, garlic, and parsley. It is a delicacy easily digested." These suggestions can perhaps be explained by the fact that the two scientists confused the livid pinkgill with E. lividoalbum, reported from North America and the UK, a morphologically similar species that had not yet been identified at the time.*

Every year the livid pinkgill, reported from North America and the UK, causes poisonings that, while not deadly, are nonetheless fairly serious and always very painful. Known as resinoid syndrome, this poisoning presents with nausea, vomiting, diarrhea, and violent intestinal pain. This mushroom can be recognized, thankfully, by its rather massive stature, its gills that are yellow and then pink (the spores are pink and stain the gills when fully mature), and its thick flesh that gives off a flour-like odor. *Entoloma lividoalbum*, reported from the UK and North America, has a similar appearance but is less fleshy, its cap becomes significantly paler as it dries (hygrophanous), and its gills are initially white, not yellow It is reported as toxic in North America.

Warning

The livid pinkgill seems to be confused most often with the clouded agaric (p. 61), in spite of the latter's crowded and decurrent gills that are never yellow or pink and its flesh's lack of a flour-like odor.

Did you know?

There are many *Entoloma* species (around 1000 worlwide) and they are morphologically quite diverse. Most beginners have a difficult time believing that mushrooms as different as the livid pinkgill and the mousepee pinkgill (*E. incanum*)—a small, skinny mushroom reported from UK and North America and known for its magnificent green-blue colors that grows in calcareous soil on lawns—could belong to the same genus. There are large and fleshy *Entoloma*, others that are small and thin, some with a stem, some without, some that associate with tree roots (mycorrhizal), others that feed on decomposing organic matter (saprophytes), and more; there are even subterranean *Entoloma* that resemble truffles.

◄ *Mousepee pinkgill*

False
chanterelle

Hygrophoropsis aurantiaca (Wulf. : Fr.) Maire

- ⊙ **Cap:** *reaches 3 inches (8 cm) in diameter, not fleshy and soft, matte or a little velvety, yellow to orangey-yellow, sometimes whitish.*
- ⊙ **Gills:** *decurrent, crowded, forked-anastomotic, yellow to orangey-yellow, easy to detach from the flesh of the cap; spore print white to cream.*
- ⊙ **Stem:** *the same color as the cap or just about.*
- ⊙ **Flesh:** *soft, orangey-yellow; sweet flavor and faint odor.*
- ⊙ **Habitat:** *especially beneath conifers, often on wood.*
- ⊙ **When to observe:** *Late summer through autumn.*

The false chanterelle, reported from North America and the UK, is very common, particularly under conifer trees where it stains fallen needles an orangey yellow color. People often confuse it with the chanterelle (p. 113) because of its colors, but it can be easily recognized by its crowded gills: upon closer examination, these gills are irregular, forked, and connected to each other by lateral folds, which is why they are said to be "anastomotic." They can also be easily separated from the flesh of the cap with a fingernail or the tip of a knife, a reminder of the connection between the false chanterelle and the boletes (bolete tubes also "peel" easily).

Did you know?

In many field guides the false chanterelle is considered an edible mushroom. The fact remains, however, that it can be confused with the toxic jack-o'-lantern mushroom (p. 199) and its soft flesh is as odorless as it is tasteless. Gastric problems have been reported after consumption of the false chanterelle. It is better to harvest yellow-leg chanterelles (p. 119) growing in the same habitat because they are far better edibles.

Sulphur
tuft

Hypholoma fasciculare (Huds. : Fr.) P. Kumm.

- ◉ **Cap:** *reaches 2.5 inches (6 cm) in diameter, red-dish-brown to bright greenish-lemon yellow.*
- ◉ **Gills:** *adnate to emarginate, typically greenish then greenish-gray; spore print brown-purple to gray-purple.*
- ◉ **Stem:** *yellow to ochre-yellow with a purplish-black ring-shaped zone.*
- ◉ **Flesh:** *yellow; very bitter taste and faint odor.*
- ◉ **Habitat:** *in tufts beneath deciduous and coniferous trees.*
- ◉ **When to observe:** *All year long except in freezing temperatures.*

The sulphur tuft, reported from North America and the UK, is probably the most common of the mushrooms growing in bouquets on dead wood. It is highy toxic, but its bitter flesh ensures that it is very rarely consumed. It appears just about all year long and is difficult to confuse with other mushrooms because of its characteristic greenish gills. The brick tuft (*H. lateritium*), reported from North America and the UK and found on deciduous wood, looks somewhat similar, but its cap is a bright red-orange "brick" color, its gills are not greenish or only slightly, and its flesh is only mildly bitter. A much rarer mushroom, the conifer tuft (*H. capnoides*), reported from the same geographic range, is sweet, has whitish to yellow gills maturing to grayish-purple, and grows on conifers in mountainous areas and colder regions.

Did you know?

Cortinarius species are well-known for their unique ring, what mycologists refer to as a cortina. It is only visible in young specimens and is made up of fine filaments that make it look like a spider's web. It would be easier to remember if the cortina only existed in *Cortinarius* mushrooms, but no such luck: developed *Hypholoma* specimens, for example, have thin filaments on their stems that form a fine and delicate blackish line, colored by their spores, their own version of a cortina.

False
saffron milkcap

Lactarius deterrimus R. Heim & Leclair

- ◉ **Cap:** up to 5 inches (12 cm) wide, orange then turning green without distinct concentric circles.
- ◉ **Gills:** somewhat decurrent, salmon orange then stained with green; spore print ochre-cream.
- ◉ **Stem:** bright orange without darker indentations, sometimes with a white band just below the gills.
- ◉ **Flesh:** orange with reddish-orange milk; tangy or bitter flavor and faint odor.
- ◉ **Habitat:** only under spruce and fir trees.
- ◉ **When to observe:** Mid-summer through autumn.

The false saffron milkcap is often confused with the saffron milkcap (p. 75), which can be identified not only by its exclusive association with pine trees, but also by its cap, usually marked with concentric circles, and its stem, which has dark dimples. While the false saffron milkcap is not poisonous, its somewhat peppery or bitter flesh makes it a mushroom not recommended for the kitchen. A lookalike, *Lactarius salmonicolor* (p. 192), reported from North America and the UK, grows under fir trees or in mixed forests, and is a brighter salmon orange with a flavor that is just as unpleasant.

Warning

If you are in the habit of buying fresh mushrooms at the market, make sure you are not tricked into inadvertently buying *Lactarius salmonicolor* or false saffron milkcaps being sold as saffron milkcaps (and at saffron milkcap prices), as sometimes happens. Note that *Lactarius salmonicolor* very rarely turns green (saffron milkcaps always have at least a few green markings) and that the false saffron milkcap does not have concentric circles on its cap. And finally, learn to identify the fir and spruce needles that may be in the crates with the substitute mushrooms: they are smaller and quite unlike the pine needles that would be found in the saffron milkcap's habitat.

Ugly
milkcap

Lactarius necator (Bull. : Fr.) Pers.

- ◉ **Cap:** *reaches 6 inches (15 cm) in diameter, olive to blackish-brown with olive-yellow hues along the edge.*
- ◉ **Gills:** *adnate or somewhat decurrent, whitish then greenish-white, olive-colored in bruised areas; spore print cream.*
- ◉ **Stem:** *greenish-yellow with a few darker indentations.*
- ◉ **Flesh:** *whitish with white milk that dries a greenish color on the gills; very peppery flavor and faint odor.*
- ◉ **Habitat:** *especially under birch and spruce trees.*
- ◉ **When to observe:** *Autumn.*

The *necator* part of this mushroom's Latin name means "murderer," which should say something about its edibility. The sharp taste of its flesh makes it inedible, but it also contains a carcinogenic substance called necatorin that should dissuade even those palates that are fond of spicy dishes. If you have sodium hydroxide (a substance sometimes used to unclog pipes) on hand, you can perform the following test: place a drop of this solution on the cap of the ugly milkcap and you will see it turn a beautiful dark purplish-blue color. This reaction is unique to European *Lactarius* species and a characteristic feature. *Lactarius necator* is not reported from the UK, and rarely from North America.

Did you know?

The French common name "Agaric meurtrier" ("murderer *Agaricus*") that would serve as the basis for this mushroom's current Latin name was first invented by French mycologist Jean-Baptiste Bulliard in his superb publication *Herbier de la France*, or *Collection complète des plantes indigènes de ce royaume* ("Herbarium of France," or "Complete collection of plants native to this kingdom"). This text was composed of nine volumes and six hundred colored panels. Curiously, one need only look at Bulliard's original panels to notice that the Agaric meurtrier does not at all resemble the ugly milkcap and is almost certainly the woolly milkcap (p. 193).

But, the happenstance of history and the demands of nomenclature were such that *Lactarius necator*, a name Bulliard originally crafted for another species, is now the only Latin name used for the ugly milkcap.

Lactarius
salmonicolor

Gröger

- ◉ **Cap:** reaches 5 inches (12 cm) in diameter, a uniform bright orange color, very few green areas or concentric zones.
- ◉ **Gills:** somewhat decurrent, bright orangey-yellow more or less salmon-colored; spore print pinkish-buff.
- ◉ **Stem:** bright orange, with or without darker indentations.
- ◉ **Flesh:** orange with bright orange milk; unpleasant flavor, often astringent or peppery.
- ◉ **Habitat:** only under fir trees.
- ◉ **When to observe:** Autumn.

Lactarius salmonicolor, reported from eastern North America and the UK, is common in colder or mountainous regions where its only host, the fir tree, grows. It is a brighter orange color than other Lactarius in the group (see saffron milkcap, p. 75, and false saffron milkcap, p. 190) and does not have the same tendency to turn green, which usually makes it easier to recognize. While it is not really poisonous, consuming it is not recommended because the bitterness of its already unpleasant flesh can irritate the stomach and intestines of sensitive individuals. To avoid making a mistake, just remember that in the group of Lactarius with orange milk, the only good edibles grow under pine trees.

Warning

When we attempt to identify mushrooms outside of their natural habitat—when they have been harvested and placed in a basket or crate, for example, and if the person gathering mushrooms did not make a note of the specimen's surroundings during the harvest—a lack of environmental information can be a serious impediment. In these situations, it is crucial to look for clues such as small leaf fragments or needles, especially at the base of the mushroom's stem, that can guide us toward a solution.

Woolly
milkcap

Lactarius torminosus (Schaeff. : Fr.) Pers.

- ⊙ **Cap:** *reaching 6 inches (15 cm) in diameter, pinki-sh-cream to salmon orangey-red, bearded and furry particularly around the edge, typically with distinct concentric circles.*
- ⊙ **Gills:** *somewhat decurrent, cream with pinkish sheen; spore print pale cream.*
- ⊙ **Stem:** *whitish or pinkish, often with a pink band at the top.*
- ⊙ **Flesh:** *whitish or pinkish; white milk.*
- ⊙ **Habitat:** *only under birch trees.*
- ⊙ **When to observe:** *Late summer but especially in autumn.*

The woolly milkcap, reported from North America and the UK, can cause nausea, vomiting, and diarrhea, if consumed raw. It only associates with birch trees and its bearded, pinkish cap makes it very easy to recognize. It can only be confused with one other *Lactarius* that is just as inedible, the bearded milkcap (*L. pubescens*), which grows in association with the same tree and is reported from the same geographic area. The bearded milkcap is smaller and paler and its cap never has concentric circles.

Did you know?

Other *Lactarius* species have a bearded cap border like the woolly milkcap, but they are much rarer and are all inedible. The whiskery milkcap (*L. mairei*) grows under oak trees in warm regions of southern Europe and has a remarkably spiky bearded cap. *Lactarius tesquorum* strongly resembles the woolly milkcap, but it is smaller (its cap rarely reaches 2.75 inches in width) and grows only under sun roses, small shrubs in southern European Atlantic and Mediterranean regions.

- ◉ **Cap:** *reaches 12 inches (30 cm) in diameter, sunken in the center, matte and velvety, white or whitish with many or few rusty brown markings.*
- ◉ **Gills:** *somewhat decurrent, not very crowded, whitish or cream; spore print cream.*
- ◉ **Stem:** *white then stained ochre.*
- ◉ **Flesh:** *white with white milk; the flavor of the flesh is very peppery, but the milk is sweet, and the odor is somewhat acidic.*
- ◉ **Habitat:** *under deciduous trees.*
- ◉ **When to observe:** *Summer and autumn.*

Fleecy
milkcap

Lactarius vellereus (Fr. : Fr.) Fr.

The very common fleecy milkcap, reported from the UK and North America, is easy to recognize: its impressive size makes it difficult to overlook, and a glance at the velvety surface of the cap is enough to identify it. This is the mushroom that leaves behind huge holes filled with water and leaves at the end of the season. Its very peppery flesh makes it inedible, though it is used dry and in powder form as a pepper substitute in some eastern European countries.

Other species

The fleecy milkcap is part of a group of mid-size to large white *Lactarius* that also includes the peppery milkcap (*L. piperatus*), reported from the same geographic area, which is less common, grows in the summer, and can be distinguished by its smaller size, its matte cap that is not velvety, and its crowded gills. *Lactarius bertillonii*, reported from the UK but not North America, has a felty cap like the fleecy milkcap, but its milk is as tangy as its flesh. None of these white *Lactarius* are edible because their flavor is always much too peppery.

Did you know?

In Europe and the rest of the Northern Hemisphere, the distinction between *Russula* and *Lactarius*—the two mushroom genera with grainy flesh (p. 99)—is usually easy to see. Not only do *Lactarius* release a milk when broken, but their cap is also often marked with concentric circles and is roughly the same color as the gills and stem. *Russula* often have pale stems and gills, creating a contrast with their colorful caps. Mycologists traveling to tropical countries are utterly disillusioned when they encounter authentic *Russula* species that are all one color with zoning on the cap and *Lactarius* species without milk!

Astonishing landscapes under the microscope...

The various features visible to the naked eye or with the use of a magnifying glass are sufficient to identify a large number of mushrooms, especially with a little experience that comes from spending time with quality mycologists. Very quickly, however, the microscope and the characteristics it is able to unveil become indispensable. In Lactarius and Russula, for example, mycologists attentively observe the spores. While usually spherical, these spores present warts on their surface whose shape, size, and arrangement are often a precious source of information. Mycologists also study the cells that make up the surface of the cap, which often have their own characteristic shape and arrangement.

- **Cap:** *up to 1.5 inches (4 cm) wide, striated along the edge, highly variable in color, often lilac or bluish-pink, paling significantly as it dries out.*
- **Gills:** *adnate to emarginate, rather crowded, white to pale lilac-gray; spore print white.*
- **Stem:** *smooth and somewhat translucent, pale at the top and the same color as the cap going toward the base.*
- **Flesh:** *thin; sweet flavor and strong radish odor.*
- **Habitat:** *under deciduous and coniferous trees.*
- **When to observe:** *Summer and autumn.*

Lilac
bonnet

Mycena pura (Pers. : Fr.) P. Kumm.

The lilac bonnet is a common mushroom, reported from North America and the UK, that is highly toxic and easy to confuse with *Laccaria* species—especially when it dries and becomes very pale, resembling a dehydrated amethyst deceiver (p. 74)—so it is important to know how to recognize it. It has pale and crowded gills and its flesh has a distinct radish-like odor that is noticeable if you smell the gills after rubbing them your finger.

Other species

There are other *Mycena* with a flat cap and a radish odor, the most common of which is the rosy bonnet (p. 198), which can be recognized by its color. The blackedge bonnet (*M. pelianthina*), reported from the same geographic regions, has much darker colors and can easily go unnoticed on the forest floor: its cap is a dark purplish-gray, and its pale lilac-gray gills have a characteristic dark purplish-brown edge. All of these species are very poisonous.

Did you know?

Current scientific knowledge estimates that there are probably more than 250 *Mycena* species in Europe, and many new species are described in mycological journals every year. The reason all of these species have not been discovered before now is that identifying *Mycena* is essentially based on microscopic evidence, requiring meticulous and patient study in a laboratory and careful examination of the spores and other cells. Perhaps the majority of mycologists are discouraged by the enormity of the task, and maybe this is why more of them are not choosing to study *Mycena*.

Mycena: a mixed group

Mycologists still have a rather broad view of genus Mycena, and include within it mushrooms that are scrawny and slender, mushrooms that have conical caps like the snapping bonnet (M. vitilis) or the clustered bonnet (M. inclinata)—both of which are reported from North America and the UK and are nontoxic without being totally edible—and extremely toxic mushrooms that are fleshier with flat caps, like the lilac bonnet (p. 197) and the rosy bonnet (p. 198). It has been demonstrated, however, that these mushrooms are not actually related to each other. Species related to the lilac bonnet form a very coherent group that is distinct from other Mycena, and some specialists already classify them in their own genus, genus Prunulus. In the future we may have to get used to names like Prunulus purus, Prunulus roseus, etc.

Rosy
bonnet

Mycena rosea (Bull.) Gramberg

- **⊙ Cap:** *reaches 2.5 inches (6 cm) in diameter, striated along the edge, a beautiful bright pink, becoming significantly paler as it dries out.*
- **⊙ Gills:** *adnate to emarginate, fairly crowded, pale pink; spore print white.*
- **⊙ Stem:** *smooth and a little translucent, often club-shaped, pinkish-white.*
- **⊙ Flesh:** *thin; sweet flavor and strong radish odor.*
- **⊙ Habitat:** *under deciduous and coniferous trees.*
- **⊙ When to observe:** *Summer and autumn.*

The rosy bonnet, reported from North America and the UK, is closely related to the lilac bonnet (p. 197) and is just as toxic. It is therefore essential that it not be confused with *Laccaria* species (p. 74), so keep in mind that none of the *Laccaria* smell like radishes. The rosy bonnet can be distinguished by its robust stature (for a *Mycena*), its pink cap that is striated along the edge and loses its color, and, of course, its characteristic odor. Its pale pinkish stem is often enlarged at the base. Some rosy bonnets are completely white, and these are members of the *candida* form.

Did you know?

The scent of turnip, radish, or potato is a very common odor among mushrooms. It is found in *Mycena*, of course, in *Cortinarius* and *Amanita* (see the false death cap, p. 169, for example), and it is the most widespread odor in genus *Hebeloma*, a group of mushrooms related to *Cortinarius* that have a brownish or café au lait spore print instead of the rust-colored print typical of *Cortinarius* species. On the other hand, none of the *Tricholoma* smell like turnips (or if they do, only mildly, and the odor is blended with a more flour-like scent), and almost no *Lepiota* or *Hygrophorus* mushrooms carry this odor. Scents are extremely important features in mushroom recognition.

Jack-o'-lantern
mushroom

Omphalotus illudens (Schwein.) Bresinsky & Besl

- ⊙ **Cap:** *reaches 6–8 inches (15–20 cm) in diameter, funnel-shaped, smooth, bright orange to orangey-red.*
- ⊙ **Gills:** *decurrent, crowded, orangey-yellow to bright orange; spore print white.*
- ⊙ **Stem:** *fibrous, the same color as the gills.*
- ⊙ **Flesh:** *orangey-yellow; sweet flavor and pleasant odor.*
- ⊙ **Habitat:** *usually in tufts on logs and roots of deciduous trees (rarely on conifers).*
- ⊙ **When to observe:** *Summer and autumn.*

This is a mushroom you should know how to recognize: reported from North America and the UK, it is still all too often confused with the chanterelle (p. 113), and the poisonings it is responsible for can be fairly serious, particularly in elderly or weakened individuals. It should be noted that the jack-o'-lantern takes pleasure in not matching its description. While it is typically a large mushroom that grows in tufts on dead wood, it is sometimes found isolated and appearing to grow from the ground (it is actually associating with tree roots). Confusion in this case is natural, so be sure to note that the jack-o'-lantern has true gills, not separated folds like the chanterelle, and that it usually has a distinctly brighter color.

Did you know?

Reading the common name for this mushroom, you may be wondering how it resembles a grinning pumpkin glowing on Halloween, but the jack-o'-lantern mushroom does indeed emit light, however faintly. To observe this phenomenon, just look at the gills of mature specimens in almost total darkness. Bioluminescence (the natural emission of light by living organisms) can be seen in around fifty mushrooms including the luminescent panellus (*Panellus stipticus*), a small mushroom that is quite common on dead wood, or honey fungus mycelium (p. 58).

Pale brittlestem

Psathyrella candolleana (Fr. : Fr.) Maire

- ◉ **Cap:** *up to 2.5 inches (6 cm) wide, dry and matte, delicate, ochre then almost white as it grows or dries out, marked with fine white or brownish labile scales.*
- ◉ **Gills:** *adnate, whitish then pale purplish-gray and finally brown; spore print dark purplish-brown.*
- ◉ **Stem:** *whitish.*
- ◉ **Flesh:** *thin; sweet flavor and faint odor.*
- ◉ **Habitat:** *in tufts or isolated, in grass or forests.*
- ◉ **When to observe:** *Spring through autumn.*

Genus *Psathyrella* is very complex. There are probably over 200 species in Europe and telling them apart is usually done with the help of a microscope. Luckily, some of them are exceptions to this rule and can be recognized with the naked eye like the pale brittlestem. This fragile little mushroom is very common and often grows on lawns, has a characteristic pale cap with fine brownish scales (sometimes only visible under a magnifying glass), and a white stem that contrasts with its gray-brown gills. Several closely related species exist but distinguishing each one is truly a task for the experts. *P. candolleana* is commonly reported across North America and the UK.

Did you know?

Psathyrella are currently the subject of intensive study: this genus was once considered to be relatively homogenous, but specialists are now having difficulty separating its members from genus *Coprinus* (pp. 69, 182, 183), a group whose definition has also undergone profound modifications in recent years. There is little doubt that a few years from now, many *Psathyrella* will find a new home with the *Coprinus* and new genera will probably be created. The progress of scientific knowledge often takes place with small revolutions like this one.

Shaggy
scalycap

Pholiota squarrosa (Weigel : Fr.) P. Kumm.

- ⦿ **Cap:** *reaches 6 inches (15 cm) in width, dry, completely covered with shaggy upturned scales, brown against an ochre background.*
- ⦿ **Gills:** *adnate, yellowish then rust brown; spore print rusty brown.*
- ⦿ **Stem:** *fibrous and dry, covered with the same scales as the cap under a well-formed fibrillose ring.*
- ⦿ **Flesh:** *cream; sweet flavor and faint odor of garlic or onion.*
- ⦿ **Habitat:** *in tufts on various deciduous or coniferous trees.*
- ⦿ **When to observe:** *Summer and autumn.*

This beautiful species, reported from North America and the UK, is not very common, but easily recognized by its cap and stem that are covered in upturned brown scales. It should also be noted that it is not at all viscous, which should help distinguish it from *Pholiota jahnii*, reported from the UK but not from North America, which has a cap that is not only an orangey-yellow color but also gluey to the touch. *P. squarrosoides*, much rarer (reported from North America but not the UK) resembles the shaggy scalycap but is paler and has a viscous layer under its scales. There are other *Pholiota* that are often hard to tell apart, but many of them are distinctly viscous.

Did you know?

Pholiota species are among the hundreds of mushrooms that, without our even realizing it, digest tons of dead wood that would accumulate year after year in our forests were it not for these natural trash collectors. The majority of them go totally unnoticed as their mycelium discreetly digests the long molecules of lignin and cellulose that form the wood. It is estimated that mushrooms transform up to 100 tons of dead wood per hectare (forty tons per acre) per year into humus.

- ⊙ **Cap:** reaches 6–8 inches (15–20 cm) in diameter, matte, can take on every color from yellow-green to dark purple, often with concentric wrinkles along the edge.
- ⊙ **Gills:** rather crowded, cream then yellow with edges that are lemon yellow or tinted pink; spore print yellow-orange.
- ⊙ **Stem:** often club-shaped, brittle, typically pale but with a pink-red band just beneath the gills.
- ⊙ **Flesh:** fairly thick, firm, and brittle; sweet flavor and faint odor.
- ⊙ **Habitat:** under deciduous and coniferous trees.
- ⊙ **When to observe:** Summer through early autumn.

A   B   C   D

▲ *Various colors of the olive brittlegill*

Olive
brittlegill

Russula olivacea (Schaeff. : Fr.) Pers.

The olive brittlegill, reported from the UK but rarely from North America, is an enormous *Russula* with sweet flesh and gills that take on a lovely dark yellow hue in adult specimens. It should be noted that the gills are pale in young specimens but change color as the spores mature. Its cap is variable in color and often wrinkled or stretched along the edge. In addition, its typically massive stem usually possesses a pink band just below the gills. It often grows beneath beech trees but can also be found under spruces in mountainous areas of Europe.

Other species

Russula are not easy to tell apart for a beginner in mycology, and the olive brittlegill has many lookalikes. *R. alutacea*, for example, reported from North America and the UK, has a cap the same color as its gills, but is usually smaller and its stem is pink at the base. The crab brittlegill (*R. xerampelina*), reported from North America and the UK, grows under pine trees and is a bright red or winy color; its gills are never as yellow as the olive brittlegill's and its yellow-ing flesh releases a distinctive odor described as the smell of cooking shellfish.

Poisonous or not?

Up until a few years ago, the olive brittlegill was considered to be an edible mushroom, albeit one of mediocre quality. It has nevertheless caused several dozen poisonings, most notably in Italy, where it is widely consumed and sometimes even sold in market stalls. Fortunately, the symptoms it causes are not serious and are consist of gastrointestinal irritation and vomiting. Many believe it is only indigestible when it has not been properly cooked, but it is still best to consider it a potentially poisonous mushroom and stop consuming it.

Did you know?

If your mushroom field guides are a bit out of date, you may read that all *Russula* with sweet flesh are edible. At one time, whether or not the flesh tasted bad (either too peppery or too bitter) was the only criterion for inedibility in this group. It was the olive brittlegill that demonstrated that certain sweet *Russula* are in fact indigestible, a discovery that sowed seeds of doubt in the minds of even the most dedicated mushroom eaters. While we wait to find out more, it is best to limit your *Russula* consumption to those that are widely known to be good edibles like the ones described in this book (pp. 93–99).

Sickener

Russula emetica (Schaeff. : Fr.) Pers.

- ◉ **Cap:** flat, a beautiful brilliant red, somewhat ribbed along the edge.
- ◉ **Gills:** white with a vaguely blue-green sheen; spore print white.
- ◉ **Stem:** white, often somewhat club-shaped.
- ◉ **Flesh:** white; very acrid and pleasant odor of coconut.
- ◉ **Habitat:** under conifers and in humid soil.
- ◉ **When to observe:** Summer and autumn.

The adjective "emetic" means "causes vomiting." Just a small piece of this *Russula* is enough to trigger a persistent and unpleasant burning sensation on the tongue. It is therefore easy to imagine how, if eaten in larger quantities, this *Russula* could cause nausea. All *Russula* that belong to the sickener mushroom group are very peppery and give off a pleasant and characteristic odor. *Russula emetica* is reported from North America and the UK.

Closely related species

The sickener is not the most common of the "emetic" *Russula*, referring to those with red caps, white gills, and peppery flesh with a "coco" smell. The most common is probably *Russula silvestris*, reported from the UK but not North America, which grows under oak trees and displays a pastel pink-red cap that quickly loses its color to turn pale pink. *Russula silvicola*, reported from North America, is very similar to *R. emetica*, but occurs in upland habitats. *Russula fageticola*, reported to be a synonym of *R. nobilis* and reported from North America and the UK, on the other hand, grows under beech trees and its cap is often a mixture of pink-red and cream.

Blackening
brittlegill

Russula nigricans (Schaeff. : Fr.) Pers.

- ⊙ **Cap:** *reaches 6–8 inches (15–20 cm) in width, often sunken in the center, white when young but quickly taken over by dark gray-brown and blackish coloring.*
- ⊙ **Gills:** *widely spaced, rigid and very brittle, brownish-cream, often with black edging, turn red and then black when bruised; spore print white.*
- ⊙ **Stem:** *brownish to blackish-gray, firm and brittle like a piece of chalk, turns red and then black when bruised.*
- ⊙ **Flesh:** *fairly thick, firm and brittle, turns red then black when broken; sweet flavor and unpleasant earthy or cheesy odor.*
- ⊙ **Habitat:** *under deciduous and coniferous trees.*
- ⊙ **When to observe:** *Summer and autumn.*

The blackening brittlegill is one of the more common *Russula*, reported from North America and the UK. It is difficult to mistake it for another mushroom, except for a few other blackening species, because of its widely spaced gills and its flesh that reddens and then turns black when exposed to air. It is almost completely rotproof and leaves black "cadavers" strewn over the forest floor when winter arrives. Blackening brittlegills are sometimes invaded by other fungi that we might describe as necrophages: the powdery piggyback (p. 255), for example, forms small colonies on old *Russula* specimens and can be recognized by the thick layer of brownish dust on its cap.

Closely related species

Other *Russula* species have flesh that turns black when exposed to air, and one of the most common, reported from North America and the UK, is the crowded brittlegill (*R. densifolia*), a small species with a cap that barely reaches 3 inches (8 cm) in diameter and is often pale around the edge with crowded gills and flesh that is usually sweet or a bit peppery. *R. adusta*, reported from North America and the UK, has flesh that slowly turns pink and may have the odor of old wine casks. *R. acrifolia*, reported from the UK, but rarely from North America, has flesh that is nearly always peppery, fairly crowded gills (though not as crowded as *R. densifolia*) and a brown cap that is often somewhat viscous and grows to a larger size (4–5 inches wide, 10–12 cm). *Russula dissimulans*, reported from North America, has, until recently, been identified as *R. nigricans*. Good observation skills will help you distinguish between all these similar species. None of these blackening *Russula* are edible.

▲ *R. acrifolia*

Verdigris
roundhead

Stropharia aeruginosa (Schaeff. : Fr.) Pers.

- ⊙ **Cap:** *reaches 2.75 inches (7 cm) in diameter, viscous, a beautiful blue-green that does not lose its color except in the center, which becomes ochre-colored as it ages, with an abundant crown of white flakes around the edge.*
- ⊙ **Gills:** *emarginate, gray-brown with a lilac sheen then dark purplish-gray with paler whitish edging; spore print grayish-violet.*
- ⊙ **Stem:** *bluish, covered in pale flaky scales beneath a fairly well-formed ring whose upper face is colored dark purplish-gray by the spores.*
- ⊙ **Flesh:** *whitish or bluish; sweet flavor and faint odor.*
- ⊙ **Habitat:** *on woody debris in and around forests, also in the mountains.*
- ⊙ **When to observe:** *Just about all year long, except in freezing temperatures.*

The verdigris roundhead, reported from the UK but rarely from North America, is easily recognizable because of its beautiful colors, its viscous cap, and its dark gills. It is very often confused with the *S. cyanea* (reported from the UK but rarely from North America) which is just as toxic and more common. *S. cyanea* has a paler blue cap that loses color more easily, fewer flakes along the edge of the cap, a ring-shaped zone instead of an actual ring, and gills that are a café au lait color with edging the same color. There is not a lot known about *Stropharia* toxicity, but they are close relatives of the *Psilocybe* mushrooms, which are sometimes extremely poisonous and even deadly.

▼ *S. cyanea*

Warning

The aniseed toadstool (p. 60), a good edible, also has green or green-blue coloring, but this is the only feature it shares with the *Stropharia* described here. Its cap is not viscous or flaky, its gills are paler, its spore print white tinged with pink, and its flesh gives off a strong aniseed odor.

Garland
roundhead

Stropharia coronilla (Bull. : Fr.) Quélet

- ◉ **Cap:** *reaches 2 inches (5 cm) in width, viscous at first then dry, light yellow to bright ochre-yellow.*
- ◉ **Gills:** *emarginate, lilac-gray; spore print dark brown-purple to black.*
- ◉ **Stem:** *white with a small membranous ring whose upper face is striated and often colored dark purplish gray by the spores.*
- ◉ **Flesh:** *whitish; sweet flavor and faint odor.*
- ◉ **Habitat:** *in grass and moss on dry lawns, in calcareous or sandy soil.*
- ◉ **When to observe:** *Especially in autumn.*

This small and elegant *Stropharia* is not often reported from North America or the UK but is found in grass on dry soil that is just beginning to grow. With its yellow cap, its lilac-gray gills, and its small striated ring, it is not difficult to recognize. When they fall from the gills, mature spores sometimes land on the ring and stain it with their color, which is another good way to recognize the garland roundhead. Less observant collectors may think it an *Agaricus* sp. There are other small ochre-yellow mushrooms that grow alongside the garland roundhead on dry lawns and none of them are edible. *Agrocybe vervacti* is one such mushroom, but it is smaller than the garland roundhead with brownish café au lait gills, a dark red-brown spore print, and a yellowish stem without a ring. It is reported from the UK, but rarely from North America.

Did you know?

Not all *Stropharia* are poisonous. The wine roundhead (*S. rugosoannulata*) is an enormous species native to North America (also reported from the UK) that has a large and heavily striated ring on its whitish stem and a purplish-brown to yellow-brown cap that can grow more than 8 inches (20 cm) wide. It is rarely found in nature in mainland Europe, though certain horticultural practices—like the dumping of hardwood mulch in the mountains—could help it become "naturalized".

▲ *Agrocybe vervacti*

▲ *Garland roundhead*

▲ *Wine roundhead*

▲ Var. filamentosum

- ◉ **Cap:** reaches 6–8 inches (15–20 cm) wide, dry, covered in gray-brown to blackish-gray scales on a gray to ochre gray background.
- ◉ **Gills:** emarginate, fairly crowded, whitish with blue-green sheen; spore print white.
- ◉ **Stem:** broad, white to cream.
- ◉ **Flesh:** rather thick, white; sweet flavor and impure flour-like odor, somewhat cloying.
- ◉ **Habitat:** especially under beech and spruce trees in mountains areas.
- ◉ **When to observe:** Summer and autumn.

Tiger
tricholoma

Tricholoma pardinum (Pers.) Quélet

This large *Tricholoma*'s friendly and appetizing appearance often causes it to be harvested in place of edible gray *Tricholoma* like the coalman (p. 105) and the gray knight (p. 106), leading to poisonings every year in Europe. This mushroom, reported from western North America but not from the UK, is rather mischievous and sometimes grows with a simple fibrillose cap (its *filamentosum* variety), without a single one of the scales that lend it its characteristic "tiger" look. To track it down, just look for its impressive stature and the pale bluish-green coloring that can be found on its gills.

Other species

The tiger tricholoma can be confused with other poisonous *Tricholoma*. A characteristic they all share is unpleasant-tasting flesh that is peppery, bitter, or both. The ashen knight (*T. virgatum*), reported from North America and the UK, for example, also grows under spruce trees, has a fibrillose cap, gills with black edges, and flesh with a peppery flavor after being chewed for a few moments. *Tricholoma josserandii*, rarely

Genus *Tricholoma*

This genus contains around 100 species in Europe, some of which are also found in North America and the UK, but those regions also have different species not found in Europe. The genus can be defined in the following manner: Tricholoma *are fleshy mushrooms with emarginate gills and a white spore print. Unfortunately, this definition also applies to other mushrooms that are not* Tricholoma *and can only be distinguished from them with microscopic characteristics that a beginner does not always have access to. True* Tricholoma *have smooth nonamyloid spores, meaning they do not turn blue when plunged into a reactive iodine-based solution. The* Lepista *(pp. 81–84), for example, often have tiny warts on their spores, and* Porpoloma *are "*Tricholoma*" with distinctly amyloid spores.*

reported from North America but not from the UK, is poisonous and grows under pine trees, but its cap is a bit greasy to the touch, often wrinkled, and its flesh releases a rancid flour-like odor.

▲ Ashen knight

▲ *Tricholoma josserandii*

Did you know?

All of the "true" *Tricholoma*—those in genus *Tricholoma*—have mycelia that associate with tree roots and are mycorrhizal mushrooms. Certain types of these mushrooms, however, feed on decomposing organic matter and are therefore saprophytes. This is the case with the plums and custard mushroom (*Tricholomopsis rutilans*). The plums and custard is easy to recognize thanks to its purple cap and bright yellow gills and the fact that it grows on old pine stumps. In tropical regions, genus *Macrocybe* contains huge saprophytic *Tricholoma* that sometimes have caps as wide as three feet.

White
knight

Tricholoma album (Schaeff. : Fr.) P. Kumm.

- ⊙ **Cap:** *up to 3 inches (8 cm) wide, white, whitish or ochre as it ages, matte and difficult to peel.*
- ⊙ **Gills:** *emarginate, white, not very crowded; spore print white.*
- ⊙ **Stem:** *white or with ochre or gray spots, fibrous.*
- ⊙ **Flesh:** *white; bitter and peppery flavor with a strong unpleasant odor that is sweetish and fruity or similar to gas.*
- ⊙ **Habitat:** *under deciduous trees, especially oaks.*
- ⊙ **When to observe:** *Especially in autumn.*

The white knight, reported from the UK and from North America, is relatively easy to recognize: its pale color and its strong distinctive odor—though sometimes hard to define—are its characteristic features. If its nauseating odor weren't reason enough to leave it alone, it should be noted that its bitter and peppery flesh is completely inedible. The chemical knight (*T. stiparophyllum*), reported from the UK and very rarely from North America, bears a strong resemblance but is larger and the edge of its cap is often ribbed. It usually grows under birch trees. There are other white mushrooms that could be confused with the white knight, but none of them exude the same strong gas odor.

Warning

The blue spot knight (p. 103) is a good edible *Tricholoma* that is all white, so care must be taken not to confuse it with the white knight. Keep in mind that the edible species has a silky cap that peels easily into triangular pieces, a stem often stained blue at the base, and a sweet flourlike flavor and odor.

Splitgill

Schizophyllum commune Fr. : Fr.

- ◉ **Cap:** reaches 2 inches (5 cm) in diameter, fan-shaped, felty and shaggy all over, whitish, grayish, or greenish, often with a scalloped edge.
- ◉ **Gills:** relatively widely spaced, pale brown, typically forked with edges split into gutter-like shapes; spore print white.
- ◉ **Stem:** absent.
- ◉ **Flesh:** thin, rubbery.
- ◉ **Habitat:** on dead deciduous wood.
- ◉ **When to observe:** All year long.

This harmless-looking little mushroom, reported from across North America and the UK, grows on dead wood and is easily recognized by its shaggy caps and unique gills that are split along the edge, reminding some mycologists of a rail yard seen from above. The luminescent panellus (*Panellus stipticus*), reported from the same geographic range, also forms small fans on dead wood, but its cap is not felty, its gills are not split, it has a small stem, and its unpleasant-tasting flesh becomes sticky when crushed between two fingers. Like the Jack-o'- lantern, it glows in the dark when fresh. All of the small mushrooms that grow on dead wood and have a very short or absent stem are called pleurotoid fungi by mycologists because their silhouette is similar to that of *Pleurotus* mushrooms (p. 109).

Warning

The splitgill grows in every country in the world, and at first glance appears quite innocent. It is even used as chewing gum in several African countries because of the rubbery texture of its flesh. Nevertheless, it has been shown that its miniscule spores are capable of germinating in the respiratory tract, causing serious cases bronchial pneumonia, particularly in people with weakened immune systems.

Bitter
beech bolete

Boletus (Caloboletus) calopus Pers. : Fr.

- ◉ **Cap:** *8 inches (20 cm) wide, whitish, café au lait or pale brown.*
- ◉ **Tubes:** *thin and crowded, yellow then olive-gray, turn a distinct blue; spore print olive-brown.*
- ◉ **Stem:** *typically red with yellow summit and a visible white network pattern.*
- ◉ **Flesh:** *white or yellowish, turns blue; bitter flavor and somewhat sour odor.*
- ◉ **Habitat:** *under coniferous and deciduous trees.*
- ◉ **When to observe:** *Summer and autumn.*

The bitter beech bolete, reported from the UK and from North America, is only abundant in mountainous areas in mainland Europe, and grows best in soil that is acidic or at least noncalcareous. It is known for its pale cap that contrasts with the bright red and yellow coloring on the stem. The stem is also covered by a visible white network pattern and the flesh has a distinctly bitter flavor, rounding out the portrait of this sadly inedible bolete. The rooting bolete (p. 216) can resemble the bitter beech bolete when it is young, but it is significantly larger, its stem does not have red hues (or very few), and it grows in calcareous soil. In eastern North America

it can be confused with *B. (Caloboletus) inedulis,* which is smaller with a pale whitish-gray cap, and *B. (Caloboletus) roseipes,* which grows under hemlock. *B. (Caloboletus) rubripes,* reported from western North America, is similar, but lacks a network pattern on its stem.

Warning

The bitter beech bolete should not be confused with the scarletina bolete (p. 133), a very good edible. The latter is easily identified by its deeper brown cap, its blue-staining flesh, and its stem, which does not have a network pattern and instead a few small red dots that can be

seen under a magnifying glass. If you are not sure, just bite off a small piece of the flesh from any boletes you harvest. This test is not dangerous and tasting bitter flesh may help you avoid a nasty surprise later on.

Wolf bolete

Boletus (Rubroboletus) lupinus Fr.

- ⊙ **Cap:** *reaches 8 inches (20 cm) in diameter, pale pink to pinkish-brown, sometimes simply brownish with a little bit of pink along the edge.*
- ⊙ **Tubes:** *thin and crowded, yellow then olive, red, or orange pores, turn blue; olive-brown spore print.*
- ⊙ **Stem:** *bright yellow or somewhat reddish, no network pattern.*
- ⊙ **Flesh:** *lemon yellow, turns blue; sweet flavor, unpleasant but faint odor of rubber.*
- ⊙ **Habitat:** *only under deciduous trees, especially oaks.*
- ⊙ **When to observe:** *Especially in summer into early autumn.*

This is a bolete that is not very common, rarely reported from North America, and not from the UK, but always easy to recognize. It is found after big summer storms and can be identified by its pink-hued cap, its yellow stem that has no network pattern, and the unpleasant odor of its flesh. It may be confused with the ruddy bolete (*B.* (*Suillellus*) *rhodoxanthus*), which is poisonous and has a network pattern on its stem and yellow flesh (reported rarely from the UK and not from North America). Dupain's bolete (*B.* (*Suillellus*) *dupainii*) is not edible, sometimes grows in the same areas (is not reported from North America nor the UK), and also lacks a network pattern, but its cap is shiny, bright red, and often appears lacquered. It is a rare bolete that must be protected.

▼ *Dupain's bolete*

it is smaller and less fleshy. Its stem is usually much narrower, its cap is often colored, and its flesh is a beet red color at the base of the stem. In addition, it prefers to grow in neutral soil, while *Boletus lupinus* enjoys calcareous soil and often grows in the company of the devil's bolete (p. 219).

Warning

The deceiving bolete (p. 134) resembles *Boletus lupinus* in its coloring and its stem without a network pattern, but

Lurid
bolete

Boletus (Suillellus) luridus Fr.

- ◉ **Cap:** *reaches 8 inches (20 cm) in diameter, dry and matte, highly variable in color, brownish-beige, orangey-brown to coppery brown but often with dull olive tones.*
- ◉ **Tubes:** *thin and crowded, yellow then somewhat olive, yellow pores that become orangey-red, blue-staining; spore print olive-brown.*
- ◉ **Stem:** *orangey-yellow with a very distinct red network pattern.*
- ◉ **Flesh:** *whitish or yellowish, with a red line just above the tubes and often beet red in the base of the stem; sweet flavor and faint odor.*
- ◉ **Habitat:** *under deciduous trees.*
- ◉ **When to observe:** *Summer and autumn.*

Reported from North America and the UK, the lurid bolete's cap colors are so variable that this mushroom is often confusing for amateur mushroom hunters. In order to see through its disguises, note that its stem is often red at the base and has a very apparent network pattern. Most importantly, it has a characteristic red line just above the tubes that is visible when the flesh is cut. This line is called "Bataille's line" after the mycologist who discovered it. It is true that a number of mushroom lovers consume the lurid bolete, but it is also regularly blamed for mild poisonings caused by undercooked specimens. For this reason, it is best to abstain from consuming it and, if you do want to try it, never consume it with alcohol, because this accentuates its indigestible nature.

Warning

The deceiving bolete (p. 134) looks like the lurid bolete and can only be differentiated from it by its lack of a network pattern on the stem. However, because mushrooms love to make already difficult identifications even more complicated, there are also forms of the lurid bolete without a network pattern. The best idea in this case is to look for the famous Bataille's line, which the deceiving bolete never displays.

Pretender

Boletus (Butyriboletus) pseudoregius Huber ex Estadès

- ◉ **Cap:** *reaches 6 inches (15 cm) in diameter, dry, bright brownish-pink.*
- ◉ **Tubes:** *thin and crowded, yellow then olive-yellow, blue-staining; spore print olive-brown.*
- ◉ **Stem:** *yellow, often with a pink band or base, a delicate but visible network pattern.*
- ◉ **Flesh:** *yellow, blue-staining; sweet flavor and pleasant odor.*
- ◉ **Habitat:** *especially under oak trees, usually in calcareous soil.*
- ◉ **When to observe:** *Summer and autumn.*

This pretty bolete, rarely reported from North America or the UK, is close to the butter bolete (p. 132) and shares many of the same features but is not always considered edible even though it is not poisonous. Its flesh is apparently not very flavorful, and since it is not a common mushroom it is better to leave it alone and allow it to reproduce in nature. Its cap is variable in color and it is not unusual to encounter specimens with an almost completely brown cap.

Warning

The royal bolete (*B. (Butyriboletus) regius*), reported from North America and the UK, resembles the pretender, but its cap is a brighter red and its flesh does not turn blue. Dupain's bolete has a somewhat viscous and stunningly red cap as well as a stem with a nearly absent network pattern. *Boletus lupinus* (p. 213) is much larger and its stem does not have a network pattern. Finally, the pale bolete (*B. (Butyriboletus) fechtneri*), reported from the UK, is a lighter-colored version of the pretender with a fine whitish bloom.

Rooting
bolete

Boletus (Caloboletus) radicans Pers. : Fr.

- ⊙ **Cap:** *reaches 8 inches (20 cm) in width, whitish, grayish-beige or putty-colored, sometimes with gray-brown or olive patches.*
- ⊙ **Tubes:** *thin and crowded, yellow then somewhat olive-colored like the pores, blue-staining; spore print olive-brown.*
- ⊙ **Stem:** *broad, often with a pointed base, yellow with a network pattern the same color, sometimes a little pink toward the base.*
- ⊙ **Flesh:** *whitish, does not turn a very bright blue; bitter flavor and unpleasant odor.*
- ⊙ **Habitat:** *under deciduous trees in calcareous soil.*
- ⊙ **When to observe:** *Summer and autumn.*

The rooting bolete, reported from the UK and from North America, is fairly common in areas with mild winters, and can be confused with the devil's bolete (p. 219) if one is only looking at the color of the cap. The rooting bolete can easily be identified, however, by its yellow pores and its stem lacking a bright pink band. Its flesh is bitter like the bitter beech bolete's (p. 212), but the latter is smaller and has a stem tinted a bright red-pink color at the base with a very apparent network pattern. The ruddy bolete (*Boletus (Suillellus) rhodoxanthus*), reported from the UK but not from North America, is another bolete with a pale cap, but its spores and stem are red, and its flesh is a particularly bright yellow.

Warning

In older field guides you will probably notice that the rooting bolete is called either *Boletus albidus* or *Boletus pachypus*. The name *Boletus radicans* has been retained as the only correct Latin name for this bolete because it is the name cited in the fundamental texts written by the father of mycology, Elias Magnus Fries. This citation is what the "Fr." after the Latin name refers to. "Pers." refers to another mycologist, Christian Persoon, who is the official creator of the *radicans* name Fries later referenced.

Oldrose
bolete

Boletus (Imperator) rhodopurpureus Smotlacha

- ⊙ **Cap:** *reaches 8 inches (20 cm) in width, typically with small dents like the skin of a toad, yellow at first then quickly turns pink, red-pink, or purplish-pink, highly blue-staining.*
- ⊙ **Tubes:** *thin and crowded, yellow, orangey or reddish-yellow pores, highly blue-staining; spore print olive-brown.*
- ⊙ **Stem:** *broad, often ventricose, bright golden yellow with a purplish-red base and a delicate red network pattern.*
- ⊙ **Flesh:** *yellow to reddish, highly blue-staining; sweet flavor and rather pleasant odor.*
- ⊙ **Habitat:** *under deciduous trees in neutral soil.*
- ⊙ **When to observe:** *Summer and autumn.*

Recognizing the oldrose bolete is not always an easy task, not just because it is not commonly reported from the UK, and even less so from North America: it is highly variable in color (in variety *xanthopurpureus* it is entirely yellow, but can also be dark purple in variety *polypurpureus*), and difficult to tell apart from closely related species that were once grouped under the collective name *Boletus purpureus*. *Boletus (Imperator) luteocupreus* (not reported from North America or the UK) for example, has a smooth copper-colored cap with a lacquered appearance, dark blood red pores in young specimens, and a stem that is dark purplish-red at the base; these distinctive characteristics are only clearly visible in young mushrooms, and identifying adult specimens is often almost impossible.

Did you know?

Boletes with red pores are divided into two groups: those that turn only somewhat blue, like the devil's bolete (p. 219), and those that turn intensely blue or blue-black when touched, like the oldrose bolete. French mycologists sometimes refer to the second group as the "salissant," or "dirty," group because it is hard to find them in perfect condition without traces of blue; the slightest bruise or even a gust of wind can alter their colors.

- ◉ **Cap:** *reaches 8–12 inches (20–30 cm) in width, pale, whitish, grayish beige or putty-colored, sometimes with gray-brown or olive patches.*
- ◉ **Tubes:** *thin and crowded, orangey-yellow or reddish pores, turn only slightly blue; spore print olive-brown.*
- ◉ **Stem:** *broad, often ventricose, typically yellow at the top and bright pink at the bottom with a fine red network pattern that is nevertheless quite apparent.*
- ◉ **Flesh:** *whitish or yellowish, turns only slightly blue; sweet flavor and unpleasant odor of spoiled meat.*
- ◉ **Habitat:** *under deciduous trees in calcareous soil.*
- ◉ **When to observe:** *Summer and autumn.*

Devil's
bolete

Boletus (Rubroboletus) satanas Lenz

The legendary devil's bolete, reported from the UK but rarely from North America, has struck fear in the heart of many a mushroom lover, but once you are familiar with this huge bolete you will wonder how it could possibly be confused with an edible bolete because there is nothing else that looks quite like it. It grows in summer and early autumn and can become very large, its cap is very pale, its pores are red or orange, its stem has a distinct network pattern, and its flesh—which exudes an unpleasant odor—turns faintly but distinctly blue. No edible bolete has all of these features. The scarletina bolete (p. 133) is often mistakenly called the devil's bolete but bears no resemblance to it: it is less fleshy, its cap is a beautiful brown, its stem has only small red spots, and its flesh turns intensely blue.

Other species

The devil's bolete can really only be confused with other inedible and poisonous boletes. The rooting bolete (p. 216) has a cap exactly the same color but its pores are yellow, its stem is never as pink, and its flesh has a bitter flavor like the bitter beech bolete (p. 212), which is smaller with a visible white network pattern. *Boletus*

A more or less toxic mushroom

The devil's bolete is certainly poisonous, but the symptoms it causes are not severe and are limited to those you might encounter with a stomach flu (nausea, vomiting, diarrhea, and abdominal pain). In some parts of Europe, most notably the Czech Republic and Slovakia, this mushroom is eaten after being cooked well. It is not entirely impossible that the devil's bolete undergoes variations in toxicity depending on where it is growing, but it is equally true that toxic molecules have been detected in its flesh and it is responsible for numerous poisonings every year. In the summer of 2011, a time of year when boletes are growing in abundance, several dozen people fell victim to the mischief of the devil's bolete after confusing it with the summer bolete.

(Rubroboletus) legaliae, reported from the UK but not from North America, has a pink or reddish-pink cap and its pale yellow flesh turns blue more intensely than the devil's bolete. The ruddy bolete (*B. (Suillellus) rhodoxanthus*) reported from the UK and not from North America, has a lovely red network pattern on its stem and its flesh is a very bright yellow.

Did you know?

Boletes form a vast and very diverse group with species all over the world. They grow almost everywhere and are particularly varied in tropical and equatorial regions, where they take on shapes and colors that are sometimes quite surprising to the eyes of a mycophile. Some boletes cause confusion when the undersides of their caps are carpeted with actual gills instead of tubes. Genus *Phylloporus*, represented in Europe by a single species, *Phylloporus pelletieri*, in North America and the UK by *P. rhodoxanthus*, and in eastern North America by *P. (Tylopilus) leucomycelinus*, brings together small gilled boletes that are very close to the red cracking bolete (p. 140).

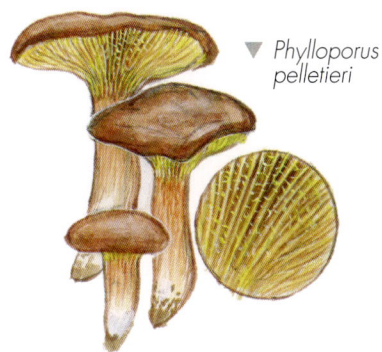

▼ *Phylloporus pelletieri*

- **Cap:** *reaches 6 inches (15 cm) in width, dry, brown to ochre or olive-beige.*
- **Tubes:** *not very thin, leading to whitish pores that turn pinkish when mature; spore print pinkish-brown.*
- **Stem:** *often club-shaped, ochre-beige and covered by a thick and prominent brown network pattern.*
- **Flesh:** *whitish or ochre-colored; very bitter flavor and pleasant odor.*
- **Habitat:** *especially under conifers.*
- **When to observe:** *Especially in autumn.*

Bitter
bolete

Tylopilus felleus (Bull. : Fr.) P. Karst.

The bitter bolete, reported from North America and the UK, is so often confused with edible boletes like the summer bolete (p. 127) and the king bolete (p. 129) that it could easily have been named the "double-crosser" instead. While you risk nothing by eating it, the unbearable bitterness from just a single mushroom is enough to render an entire dish inedible and you will have to throw away whatever you prepared with it. It is easy to distinguish the bitter bolete, though, by its thick brown network pattern that is not found on any edible boletes and by the distinctly pink pores in adult specimens. If you are uncertain, chew and spit out a small piece of the bolete you wish to identify. This harmless test will spare you from any unfortunate surprises. If you are one of the people who cannot taste bitter, you will have to carefully observe the visible distinctions listed above.

Other species

Only one *Tylopilus* species exists in Europe, but this bolete genus is heavily represented in tropical regions. The dusky bolete (*Porphyrellus porphyosporus*), reported from North America and the UK, is an inedible mushroom, with a beautiful dark brown cap and stem and gray-brown pores with a hint of purple that stain paper teal. The bitter bolete and the dusky bolete are the only two European boletes without an olive or brown spore print, and instead

▼ *Dusky bolete*

have pink and reddish-brown prints respectively, though other members of the Bolete family may have yellow or black spore prints. North America has a few Bolete species with pinkish spore prints, including *P. fumosipes*, reported from the east, and several other *Porphyrellus* and *Tylopilus* species.

Did you know?

Bitterness is not a common mushroom flavor. It sometimes appears in aging edibles like the hedgehog (p. 122), for example, but for some species it is a distinguishing feature that helps identify them. While the bitter bolete is a well-known example, we might also mention *Mycena erubescens* (reported from the UK and North America), a small species with a conical brown-gray to reddish-brown cap that grows on mossy tree trunks and has such a powerful flavor it can be sensed simply by touching the mushroom with the tip of your tongue.

Is it a mycorrhizal mushroom or not?

The bitter bolete is frequently found on old pine tree stumps, which has prompted some people to say that its mycelium does not associate with tree roots like a proper mycorrhizal mushroom and that it is really a saprotroph that feeds on dead wood. In reality, the bitter bolete takes refuge on worm-eaten logs when the weather is dry because it can find enough humidity still trapped in the softened wood. Fine tree roots are able to colonize a variety of substrates like this, and sometimes seem to take to the air: in tropical and equatorial zones, mycorrhizal mushrooms have been observed several yards in the air on the worm-eaten trunks of large trees.

Thick cup

Discina (Gyromitra) *perlata (ancilis)* (Fr. : Fr.) Fr.

- ⊙ **Mushroom** *in the shape of a cup at first, then flattens out on the ground and can grow up to 4 inches (10 cm) in width, brown to reddish-brown superior face, irregular or dented in mature specimens; paler inferior face; spore print white.*
- ⊙ **Stem:** *short but fairly developed, central.*
- ⊙ **Flesh:** *brittle, not very thick; sweet flavor and faint odor.*
- ⊙ **Habitat:** *under conifers, on mossy stumps or needles, sometimes on old campfire sites, usually in mountainous areas.*
- ⊙ **When to observe:** *Spring into early summer.*

The thick cup, reported from North America and the UK, is fairly common in mountain coniferous forests where it appears in the spring on mossy old tree stumps, and is rare in lowland areas, in the warmer areas of its range. It may be confused with the bleach cup (p. 151), which also grows in spring but does not have a separate stem, grows on bare ground in young deciduous forests, and releases a bleach-like odor. Because it is close to the false morel (p. 248), you are strongly advised not to eat the thick cup mushroom.

Warning

The orange peel (p. 150) and the bleach cup are the most well-known of the few *Peziza* species that are edible, and only the latter truly merits being included in the pan. Nevertheless, they are part of the same family as excellent edibles like morels (pp. 145–146) and truffles (pp. 155–159): the ascomycetes. This vast and varied group also includes the *Helvella* (p. 149), *Gyromitra* (p. 248), and *Xylaria* (p. 286).

Upright
coral

Ramaria stricta (Pers. : Fr.) Quélet

- ◉ **Mushroom** *in branching shrub shape, reaching 4 inches (10 cm) tall and 2.75 inches (7 cm) wide, more or less parallel branches, ochre yellow with pointy sulphur yellow tips that then become ochre; spore print golden yellow to ochre-yellow.*
- ◉ **Stem:** *thick and often short, fleshy, whitish, elongated by white mycelium threads.*
- ◉ **Flesh:** *white, thick; bitter flavor and somewhat sour odor.*
- ◉ **Habitat:** *on rotted wood from coniferous and deciduous trees.*
- ◉ **When to observe:** *Especially in autumn, but also in mid-summer in colder regions.*

This species, reported from North America and the UK, is one of the rare *Ramaria* described as being lignicolous, meaning it grows on dead wood. The majority of other species in this genus associate with tree roots through their mycelium (mycorrhizal species). This feature makes this mushroom easy to recognize, especially if we consider its specific habitat, the almost parallel positioning of its branches, the presence of white threads of mycelium, and lastly, its distinctly bitter flesh. The crown-tipped coral (*Artomyces pyxidatus*), reported from North America but not from the UK, also grows on wood, but is whitish to cream in color and the tips of its branches are flared like small trumpets.

Did you know?

All *Ramaria* have ochre spores and a spore print the same color. This is what allows us to distinguish them from genus *Clavaria* mushrooms, for example, which usually have white spore prints. This characteristic also links them to other mushrooms like the pig's ear (p. 121) that are morphologically quite different from them. The spores of the pig's ear are similar to the upright coral's, but its silhouette and the forked folds on the underside of its cap make it look more like an enormous purplish chanterelle.

▲ Crown-tipped coral

▲ Upright coral

- ◉ **Mushroom in branching shrub shape**, *reaching 6 inches (15 cm) tall and 4 inches (10 cm) wide, branches typically a beautiful pale pink with sulphur yellow tips; spore print ochre-yellow.*
- ◉ **Stem:** *thick and often short, fleshy, whitish.*
- ◉ **Flesh:** *white, thick; flavor often a little bitter and faint odor.*
- ◉ **Habitat:** *on the ground under deciduous trees, particularly beech trees.*
- ◉ **When to observe:** *Summer and autumn.*

Salmon
coral

Ramaria formosa (Pers. : Fr.) Quélet

The salmon coral, reported from North America and the UK, is not very common and is difficult to identify. Other *Ramaria* have identical coloring, so it is best to throw away any *Ramaria* with pink tones, as a rule. However, it should be noted that these pink tones tend to disappear when the mushroom ages, which increases the risk of confusing it with edible *Clavaria* like the golden coral (p. 153). The symptoms caused by salmon coral consumption are fortunately not severe and manifest between fifteen minutes and two hours after ingestion. They include abdominal pain, nausea, and diarrhea that passes within a few hours. Still, you are better off sparing yourself this rather unpleasant experience.

Other species

Ramaria neoformosa (reported from North America but not from the UK) and *R. fagetorum* (not reported from North America or the UK) grow in the same areas as the salmon coral and can only be differentiated using a microscope to study the spores as carefully as possible. The upright coral (p. 223) is also tinted pink and has yellow tips on its branches, but is usually smaller, has a slender appearance, and grows in relationship with dead wood: for this purpose, the upright coral has long white threads of mycelium at the base of its stem.

Did you know?

Lucien Quélet was a mycologist from the Jura region of eastern France, who was born in 1832. His pioneering work would serve as the foundation for the mushroom classification system that is still in use today. He described close to 200 new species, the majority of which are presented in his major work entitled *Les champignons du Jura et des Vosges* ("Mushrooms of Jura and the Vosges"). It is because of him that the salmon coral was placed in genus *Ramaria*, though another illustrious mycologist, Christian Hendrick Persoon, had originally described it as *Clavaria formosa*.

The Aphyllophorales

Before the advent of chromosome and DNA studies attempting to reconstruct the evolutionary history of living organisms, mushrooms were grouped for practical purposes in groupings based on their morphological similarities. The majority of mushrooms without gills were classified under Aphyllophorales, from the privative prefix a and the Greek phullas, meaning "leaves" or "gills," and grouped with organisms as diverse as the polypores (p. 143), Clavaria, and mushrooms that form nothing more than a crust on dead wood. Today, this highly varied group has been completely dismantled and the mushrooms that once were part of it have all been placed in groups reflecting their true biological affinities.

▲ Death caps (*Amanita phalloides*)

Deadly Mushrooms

- **Cap:** *reaches 6 inches (15 cm) in width, smooth, usually olive green or yellow-green, but sometimes all yellow or all white, typically striped with small grayish fibrils.*
- **Gills:** *free and crowded, white; spore print white.*
- **Stem:** *white with greenish-yellow variegations, a skirt-shaped ring, emerges from a large membranous sac-like volva.*
- **Flesh:** *white; sweet flavor and pleasant sweetish odor similar to wilted roses in adult specimens.*
- **Habitat:** *especially under deciduous trees.*
- **When to observe:** *Spring until late autumn.*

D e a t h
cap

Amanita phalloides (Fr. : Fr.) Link

Of the tens of thousands of mushrooms present worldwide, very few are toxic and even fewer are deadly. Unfortunately, the death cap—responsible for atleast one death a year in North America—is extremely common. Reported from the UK, it is now more commonly reported in coastal North America than in the past. It grows abundantly in a variety of different forests, and if you go for a walk at the height of mushroom season you are likely to encounter several hundred specimens at a time in some regions! It is also an elegant mushroom whose sweet flesh and delicate odor make it all too appetizing.

Warning

Luckily, the death cap does not resemble any edible mushrooms, at least not in its typically colorful form. When completely white, though, it can easily be confused with the edible *Agaricus* species (pp. 45–49) or the white dapperling (p. 85). To avoid accidentally picking the death cap or its equally deadly cousins the spring amanita (p. 230) and the destroying angel (p. 231), just remember that none of the white edible mushrooms mentioned above emerge from a membranous sac that persists at the base of the stem in mature specimens.

To cut or not to cut?

This is a question that comes up often: when harvesting mushrooms, should you carefully cut the mushroom level with the earth to leave its "root" in the ground or remove the entire mushroom? The death cap is a good example of how important it is to delicately remove the whole mushroom, especially the base of its stem. If you do not do this, you risk not seeing a critically important identifying feature: the "root," or volva. It has also been shown, at least in the case of chanterelles, that harvesting mushrooms without cutting the stem does not influence mushroom regrowth.

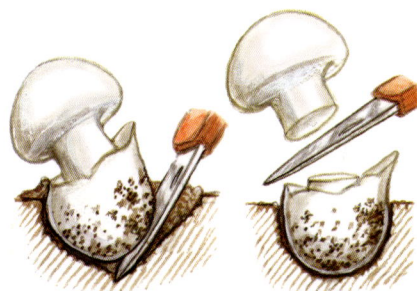

A strange doctor...

Dr. Pierre Bastien, a physician from the Vosges region of northeastern France with an enormous personality, made headlines in the early 1970s when he bravely ingested copious amounts of Amanita phalloides *in front of cameras to prove that the antidote he had developed to treat death cap poisoning worked. He even wrote a book,* J'ai dû manger des amanites mortelles *("I Had to Eat Deadly Amanita"), that was released by a well-known Parisian publisher. This antidote, though undeniably effective and very simple, must be administered almost immediately after ingestion, which is problematic since this poisoning is usually diagnosed too late.*

Spring
amanita

Amanita verna (Fr. : Fr.) Link

- ◉ **Cap:** *reaches 6 inches (15 cm) wide, smooth, regular shape, all white.*
- ◉ **Gills:** *free and crowded, white; spore print white.*
- ◉ **Stem:** *white, usually smooth with a skirt-shaped ring, emerging from a large membranous sac-like volva.*
- ◉ **Flesh:** *white; sweet flavor and pleasant sweetish odor similar to wilted roses in adult specimens.*
- ◉ **Habitat:** *especially under deciduous trees.*
- ◉ **When to observe:** *Spring and autumn.*

This *Amanita*, reported from North America and the UK, is a close relative of the death cap (p. 229) and is also deadly. Its unremarkable appearance and white coloring often encourage confusion with the white edible *Agaricus* (pp. 45–49) and the white dapperling (p. 85). The spring amanita was responsible for one of the most significant poisonings of the last century, which took place in Trévoux, a region of France, in 1911. Here is what Dr. Maurice Roch wrote about this poisoning in the 1913 edition of the *Bulletin of the Botanical Society of Geneva*: "The tragedy was caused by a plate of mushrooms in sauce that was served at lunch to regular customers of a particular restaurant. Those afraid to touch the dish were mocked to such an extent that everyone ate it. That evening at dinner, each one returned to his seat in good health and one joker exclaimed: 'If we have been poisoned, we shall all die together!' Around midnight, the first symptoms appeared." Twenty-three poisoned and between nine and eleven dead, according to the various accounts…

Did you know?

The spring amanita is often found in two forms: variety *verna*—usually in the spring, and rarer—and *virosa* variety *levipes*, which is much more common and grows in autumn. To tell them apart, mycologists use a ten percent potassium hydroxide solution: a drop of this turns variety *levipes* a bright yellow and has no effect on variety *verna*.

Destroying angel

Amanita virosa (Fr.) Bertill.

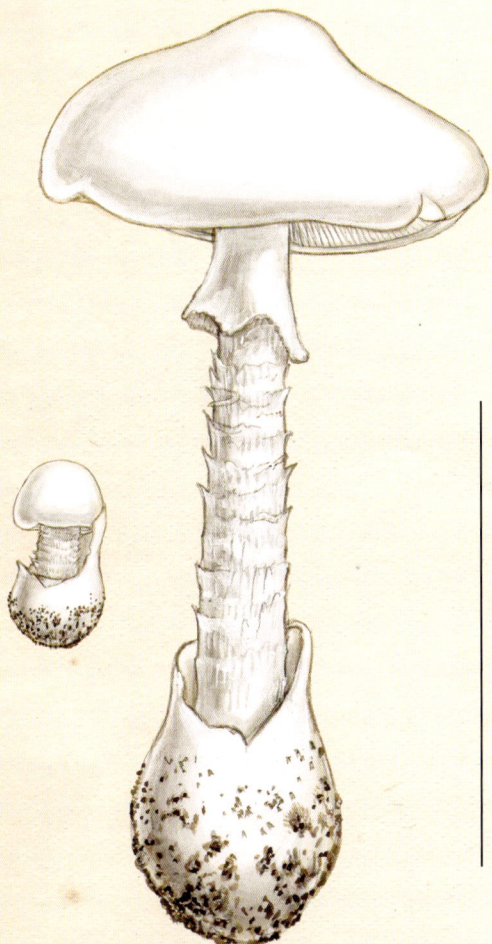

- ◉ **Cap:** *reaches 4 inches (10 cm) in width, smooth, irregular shape, tilted position on top of the stem, truncated cone shape or with a large umbo, white to cream.*
- ◉ **Gills:** *free and crowded, white; spore print white.*
- ◉ **Stem:** *white, fluffy and wispy, a sometimes "ruffled" appearance, with a fragile skirt-shaped ring that is often torn, emerging from a membranous sac-like volva.*
- ◉ **Flesh:** *white; sweet flavor and pleasant sweetish odor similar to wilted roses in adult specimens.*
- ◉ **Habitat:** *especially under deciduous trees.*
- ◉ **When to observe:** *Mid-summer through autumn.*

The destroying angel, reported from the UK and North America, is not very common in Europe as a whole, but in some regions, it is found quite often. It has a rather unique appearance, always looking somewhat poorly constructed with its irregular cap leaning to one side, and its stem more or less covered in untidy and curling scales. It is just as deadly as its cousins the death cap (p. 229) and the spring amanita (previous page) but is almost never harvested because of its odd silhouette and how rare it is. However, it has been the cause of severe poisonings and death in North America.

Did you know?

The *Amanita* have an astonishing feature: instead of growing little by little during the mushroom's development as happens in other groups, *Amanita* gills are already fully formed in young mushrooms and emerge completely fused together inside a single seamless covering. Over the course of development, space is slowly created between the individual gills; this is what mycologists describe as "schizohymenial" development.

- **Cap:** *reaches 4 inches (10 cm) wide, matte and somewhat felty, convex (without distinct umbo), uniformly dark orangey-red or reddish-brown.*
- **Gills:** *emarginate, fairly widely spaced and rather broad, dark red; spore print rusty brown.*
- **Stem:** *the same color as the cap or more yellow, fibrillose.*
- **Flesh:** *yellowish; sweet flavor and radish odor.*
- **Habitat:** *under deciduous trees, especially oaks.*
- **When to observe:** *Autumn.*

Fool's webcap

Cortinarius orellanus Fr.

The fool's webcap, reported from the United Kingdom and from northwestern North America, is not very common. It grows in soils that are not acidic, or at least noncalcareous, and dry. It only associates with deciduous trees, though conifers are the preferred hosts of its cousin, the deadly webcap (p. 234). In French, it is called the "rocou *Cortinarius*" because its coloring resembles the deep red color of this natural dye, an extract from a tropical tree called the rocouyer or achiote tree (*Bixa orellana*). The habitat of the fool's webcap, its nearly uniform coloring, its broad and widely spaced gills, and its radish-scented flesh are all excellent clues to help you recognize it.

A little bit of history

In his book Les champignons toxiques et hallucinogènes *("Toxic and Hallucinogenic Mushrooms"), renowned mycologist Roger Heim—then director of the National Museum of Natural History in Paris—explains that "the many poisonings that occurred following mushroom ingestion among Polish populations in the Konin and Alexandrov districts between 1952 and 1955 attracted significant attention. In the fall of 1952, 102 people became seriously ill and 11 died." Polish mycologists were able to identify the guilty party:* Cortinarius orellanus. *We should note that until this time, all* Cortinarius *species were considered edible . . .*

Warning

The fool's webcap is deadly: the toxic substances it produces, orellanines, cause irreparable damage to the kidneys. Fortunately, it is not responsible for very many poisonings because it does not resemble other mushrooms, at least not any edible ones. The poisonings it has caused have usually been the result of mushroom mixtures or periods of famine in nations with unstable economies like certain parts of eastern Europe.

Did you know?

Poisoning by the fool's webcap is particularly dangerous because the first symptoms appear long after ingestion. People who have been poisoned notice the first signs between twelve hours and fourteen days (three days, on average) after the fatal meal: these include vomiting, nausea, and diarrhea often accompanied by abdominal pain and loss of appetite. The renal damage, which usually evolves into chronic renal failure, appears roughly eight or nine days after ingestion. There is not yet a treatment to counteract this toxin, but deaths have become rare with the help of dialysis procedures and renal transplants.

Deadly
webcap

Cortinarius speciosissimu (rubellus) Kühner & Romagnesi

- ◉ **Cap:** *reaches 4 inches (10 cm) in width, matte and somewhat felty, typically conical or with a well-developed umbo, uniformly dark orangey-red or reddish-brown.*
- ◉ **Gills:** *emarginate, fairly widely spaced and broad, dark red; spore print rusty brown.*
- ◉ **Stem:** *the same color as the cap or more yellow, fibrillose, with more or less complete wreaths from the bright orangey-yellow veil.*
- ◉ **Flesh:** *yellowish; sweet flavor and radish odor.*
- ◉ **Habitat:** *under conifers, especially spruce, on peat soil; more frequently found in colder or mountains regions.*
- ◉ **When to observe:** *Late summer and early autumn.*

The deadly webcap, reported from the UK and North America, is a close relative of the fool's webcap and is also deadly. It can be distinguished by its cap that is more conical than convex, its stem adorned with apparent wreaths of yellow veil, and by its preference for humid conifer forests. Other morphologically similar *Cortinarius* species have been described, but not all mycologists are in agreement about whether they actually exist; they could simply be local forms of the deadly webcap or the fool's webcap.

Did you know?

The deadly webcap is part of the *Leprocybe* group of *Cortinarius* species. Members of this group are all toxic, if not deadly, and possess unique red, yellow, or olive pigments that are fluorescent when exposed to ultraviolet light. All *Leprocybe* species also have a somewhat felty cap.

Splendid
webcap

Cortinarius splendens Henry

- **Cap:** reaching 2.75 inches (7 cm) wide, distinctly viscous, bright yellow often with small reddish patches.
- **Gills:** emarginate, bright yellow then a little reddish; spore print rusty brown.
- **Stem:** distinctly marginate bulb, bright yellow, with a trace of a rust-colored cortina.
- **Flesh:** a uniformly saturated bright yellow; sweet flavor and faint odor.
- **Habitat:** under deciduous trees, especially beech.
- **When to observe:** Autumn.

"In my corner of the nation, I have a bad reputation…"[1] George Brassens could easily have written this lyric about the splendid webcap; it is an uncommon mushroom (reported from North America and the UK) that is easy to identify because of its small size, its beautiful yellow color that reaches an unusual intensity in the mushroom's flesh, and its habitat near beech trees, but it does indeed have a bad reputation. It is considered suspect and deadly in many field guides, but its toxicity has never been proven. To be on the safe side, it is best to leave it alone like the rest of its counterparts.

Did you know?

There are several other yellow *Cortinarius* with viscous caps and distinctly marginate bulbs. Telling them apart from one another is often a delicate matter, even for an expert. *Cortinarius elegantissimus*, reported from the UK and from North America, sometimes grows under beech trees as well, and is just as yellow, but is much bulkier, its cap quickly turns a rusty brown when it grows, and its flesh is yellowish and white in the cap and center of the stem.

---

[1] Translated by Pierre de Gaillande, from the song "La Mauvaise Réputation."

- **Cap:** *2 inches (5 cm) wide, smooth, often a little greasy or sticky to the touch, orangey-brown to ochre-brown, paler as it dries; the skin of the cap (cuticle) extends to form a fine transparent border along the edge.*
- **Gills:** *adnate, cream then ochre; spore print rusty brown.*
- **Stem:** *ochre, fibrillose beneath a small and fragile membranous ring.*
- **Flesh:** *thin; faint odor, a bit flour-like, flour-like flavor.*
- **Habitat:** *on dead wood from deciduous and coniferous trees, in colonies that are sometimes dense, but never tufts.*
- **When to observe:** *Summer and autumn.*

Funeral
bell

Galerina marginata (Batsch) Kühner

The funeral bell, reported from North America and the UK, is frequently encountered and grows in colonies on dead wood. It is most often recognized by its smooth cap, ochre gills, and smooth fibrillose stem with a small ring stained reddish-brown by spores falling from the gills. Along the edge of the cap, a small transparent border formed by the skin of the cap completes the portrait of a mushroom that you need to be able to identify.

A very toxic mushroom

The toxic substances produced by the funeral bell are amatoxins, the same ones found in the death cap (p. 229) and other Amanita from the same group. Experiments have shown that some populations of this Galerina contain proportionally greater amounts of amatoxins than the death cap, even though the latter is often considered the more poisonous mushroom. While it is very common, the funeral bell rarely causes poisonings, and only a dozen were recorded in the twentieth century.

Warning

The sheathed woodtuft mushroom (p. 73) is a good edible that is much sought-after by harvesters in certain regions and is also an almost perfect funeral bell lookalike. To avoid making a mistake, it is important to carefully examine every specimen: the sheathed woodtuft has a stem covered in a fine scaly "sock" (armilla) that terminates in a sort of deformed ring at the top. It also does not have a transparent border on its cap, which is an essential difference that can be helpful if the specimens have been handled and the stem is no longer in perfect condition. Finally, the sheathed woodtuft grows in tufts (the stems are united at the base), while the funeral bell develops in colonies

(the stems are not usually fused at the base).

Did you know?

The funeral bell is widespread throughout the Northern Hemisphere but has also been identified in Australia. Up until a few years ago, mycologists recognized several other "species" that were morphologically similar to the funeral bell but displayed subtle differences in habitat and cap viscosity. Studies using molecular biology to compare DNA segments have proven that these "species" were in fact various forms of the funeral bell.

Blackening
waxcap

Hygrocybe conica (Schaeff.) P. Kumm.

- ⊙ **Cap:** *1.5 inches (4 cm) wide, conical, smooth and slightly viscous, tomato red, orangey-red, or orangey-yellow, turns black with handling or age.*
- ⊙ **Gills:** *ascending and emarginate, orangey to reddish-yellow, yellowish-white when the cap is yellow, turn black; spore print white.*
- ⊙ **Stem:** *fibrillose, orangey reddish-yellow, whitish at the base, turns black.*
- ⊙ **Flesh:** *thin, turns black; faint odor and sweet flavor.*
- ⊙ **Habitat:** *in the grass of lawns, prairies, and roadsides.*
- ⊙ **When to observe:** *Spring until the beginning of winter.*

This beautiful *Hygrocybe* is commonly reported across North America and the UK and often shows its small conical stem on green lawns, where it can be seen from a distance. It blackens easily, so much so that older specimens look more like pieces of charcoal than mushrooms and can be difficult to spot. There are other red *Hygrophorus* species, but few of them turn black as intensely as the blackening waxcap. *H. pseudoconica* turns black but has a larger cap (up to 3 inches (8 cm) wide) that is more fibrillose than viscous and is now considered a synonym of *Hygrocybe conica*.

Did you know?

The blackening waxcap appears to have caused only a few poisonings in Europe, all of which have been mild, and it does not resemble any edible species. Confusion is therefore highly unlikely, and the true danger of this little mushroom is somewhat relative. Nevertheless, it has been blamed for at least one serious poisoning in China, and caution should still be taken.

▲ *The slow darkening of the blackening waxcap is an excellent way to recognize this mushroom.*

- ⊙ **Cap:** *2.75 inches (7 cm) wide, smooth and distinctly fibrillose, all white at first then brownish to ochre, reddens (pink to orangey-red).*
- ⊙ **Gills:** *emarginate, whitish then grayish-brown, turn red; spore print brown.*
- ⊙ **Stem:** *fibrillose, smooth, not bulbous or with an underdeveloped onion-shaped bulb, whitish and turns red.*
- ⊙ **Flesh:** *white; pleasant fruity or honey odor, sweet flavor.*
- ⊙ **Habitat:** *under various deciduous trees.*
- ⊙ **When to observe:** *Especially in spring in the UK (very rare in autumn); autumn in North America.*

Deadly
fibrecap

Inocybe patouillardii (*erubescens*) Bres.

The deadly fibrecap, reported from the UK but not North America, is not the only *Inocybe* found in the spring, but it is the only one with a very fibrillose cap, a smooth nonbulbous stem, and reddening flesh. *Inocybe godeyi*, reported from North America and the UK, turns red in a similar fashion, but its flesh gives off a semen-like odor and its stem is not smooth: under a magnifying glass, it appears covered by a fine white bloom, and the base of its stem has a distinctly marginate bulb. *Inocybe bresadolae*, reported from North America but not from the UK, also has a bulbous stem, is covered with a frosted bloom, and has pleasantly scented reddening flesh.

Warning

Edible mushrooms are rare in the spring. Unfortunately, the St. George's mushroom (p. 57) is white and can be confused with the deadly fibrecap when it is young: note that the deadly fibrecap has gills that are more widely spaced (they are very crowded among *Tricholoma*) and that its flesh gives off an odor that

(p. 57)

All *Inocybe* are toxic

While not all Inocybe *species have been tested, the title of this section is nevertheless probably true and deserves to be remembered. There are likely hundreds of* Inocybe *mushrooms in central Europe (many more than that in North America) and the dozens that have been studied by European chemists in detail have been confirmed as poisonous. The toxic substance responsible, muscarine, is also present in the fly agaric (p. 173), but the deadly fibrecap contains roughly 100 times more muscarine than this well-known* Amanita.

is pleasant but not at all like fresh flour. The shield pinkgill (p. 65) has a silhouette like that of an *Inocybe*, but its gills become pink as it develops, and its flesh gives off a distinct flour-like odor.

Did you know?

The great Dutch mycologist Christian Hendrik Persoon was the first person to report a poisoning caused by *Inocybe* mushrooms. In 1818, a family from Turin died after being poisoned by *Agaricus rimosus*, an old name for a group of *Inocybes* that are still not clearly described. The symptoms that are part of this poisoning became bet-ter understood over the course of the twentieth century through several heavily studied cases. They include dizziness, nausea, unusual perspiration and salivation, trembling, distorted vision that causes objects to appear larger or as brightly colored lights, and sometimes death as a result of cardiac arrest.

- ⊙ **Cap:** *2.5 inches (6 cm) wide, smooth or more often covered in small gray-pink to brown-pink scales, sometimes only pinkish-white.*
- ⊙ **Gills:** *free, white; spore print white.*
- ⊙ **Stem:** *white at the top and muddy pinkish color at the bottom, with incomplete wreaths and small reddish-brown flakes under a slightly deformed ring area.*
- ⊙ **Flesh:** *white; pleasant fruity and mandarin-like odor, sweet flavor.*
- ⊙ **Habitat:** *lawns, sometimes in sand dunes.*
- ⊙ **When to observe:** *Especially in autumn.*

Deadly
dapperling

Lepiota brunneoincarnata Chodat & C. Martin

Instead of trying to recognize the deadly dapperling (reported from the UK, but not from North America), which is difficult without a certain amount of experience, learn to identify with certainty the *Lepiota* from the deadly dapperling group, which contains several species that are all toxic and morphologically similar. These are small to mid-size mushrooms with free white gills and caps covered in small colorful scales that sit upon a whitish background. Their flesh is often fragrant but depending on the species may give off a variety of scents reminiscent of citrus or rubber, in the case of the stinking dapperling (p. 244).

in width, and their stems have a well-defined ring that slides. Be wary, not all *Lepiota* that have a ring that slides are non-toxic. Be sure not to confuse the Scotch bonnet (p. 90) with older "overripe" specimens of the deadly dapperling.

Did you know?

Mycologists place the deadly dapperling in the *Lepiota helveola* group, a name created in 1881 by Italian mycologist Giacomo Bresadola in his work entitled *Fungi tridentini* ("The Mushrooms of Trentino"). Since that time, several other *Lepiota*, all of which are toxic, have been added to this group. Funnily enough, even 130 years later no one knows exactly what this famous *Lepiota helveola* classification means, but its name continues to act as a label for this group of small poisonous *Lepiota*.

Warning

The "small and deadly" *Lepiota* are a legend in their own right. They have sent shivers down the spine of many a mushroom lover but are paradoxically rather unknown. "How can we tell them apart from edible *Lepiota*?" people often ask. First, you must remember that edible *Lepiota* are all much larger than toxic *Lepiota*: their caps easily exceed 6 inches (15 cm)

An entire world . . .

There are many small and poisonous Lepiota, *and they are not always easy to tell apart. The most common is perhaps the fatal dapperling (L. subincarnata), reported from North America and the UK, which has muted pink tones and a pleasant mandarin scent. It is easily confused with the deadly dapperling but has lighter colors and its stem does not have an apparent dark reddish-brown veil, a typical feature in the latter mushroom. The star dapperling (L. brunneolilacea), reported only from continental Europe, also bears a strong resemblance to the deadly dapperling, but its ring is often more noticeable, and it usually grows in the sand of coastal dunes.*

Stinking
dapperling

Lepiota cristata (Bolton) P. Kumm.

- ⦿ **Cap:** *2.5 inches (6 cm) wide, often covered with small red to orangey-brown scales that are sometimes white.*
- ⦿ **Gills:** *free, white; spore print white.*
- ⦿ **Stem:** *whitish to pinkish with a small skirt-shaped ring that is very fragile and often disappears or remains hanging from the edge of the cap in adult specimens.*
- ⦿ **Flesh:** *white; strong rubbery odor and sweet flavor.*
- ⦿ **Habitat:** *in ditches, along roadsides, or in thickets, often among stinging nettle and other ruderal plants.*
- ⦿ **When to observe:** *Especially summer through autumn.*

This small *Lepiota* is very common, both in the UK and North America. It is also extremely inconsistent and loves to appear in a variety of different colorations that can trip up many mushroom lovers. In addition, its ring—a characteristic feature when well-developed—is extremely fragile and often absent in more mature specimens. The stinking dapperling can nevertheless be recognized by its often-slender silhouette as well as its strong and distinct rubber odor.

Did you know?

Many small *Lepiota* are fond of soil that is rich in nitrogen compounds, and for this reason tend to grow along roadsides, in gardens, piles of debris, and other places frequented (and altered) by humans. Certain plants like the stinging nettle have similar preferences and are called "ruderal" plants, from the Latin *rudus*, "debris." This is why mycologists who study these *Lepiota* have such great appreciation for the stinging nettle!

Brown
rollrim

Paxillus involutus (Batsch) Fr.

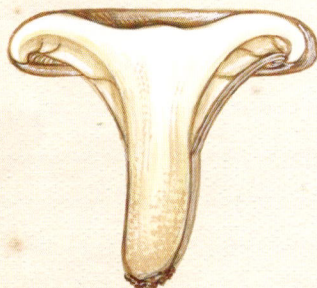

- ◉ **Cap:** *8 inches (20 cm) wide, funnel shape with the edge rolled under for some time, dark ochre-brown with olive or reddish-brown sheen, often with concentric circles.*
- ◉ **Gills:** *decurrent, easily detach from the flesh of the cap, ochre-yellow turning red in bruised areas; spore print yellow-brown to rusty brown.*
- ◉ **Stem:** *short, fibrous, reddish-brown.*
- ◉ **Flesh:** *brownish; faint fruity odor and sweet flavor.*
- ◉ **Habitat:** *often under birch trees, but under other trees as well.*
- ◉ **When to observe:** *Especially in autumn.*

Like the false morel (p. 248) and the yellow knight (p. 247), the brown rollrim, reported across North America and the UK, is a mushroom that was consumed for decades, and is sometimes still eaten, even though it is now considered deadly. Scientists have discovered that this mushroom is capable of provoking a violent anaphylactic reaction that causes the destruction of red blood cells. This reaction is followed by a variety of symptoms including nausea, vomiting, and gastrointestinal pain that can be accompanied by potentially fatal renal failure.

Did you know?

Being a mycologist is a dangerous career at times. We owe the first proof of the brown rollrim's toxicity to a very famous mycologist named Julius Schäffer who died after consuming a plate of these mushrooms in October 1944. We have since identified the molecule responsible for this toxicity as well as other potentially carcinogenic and mutagenic (causing mutations) substances. In spite of all of this, as late as the 1990s quite a few field guides were still saying that the brown rollrim was a good edible . . .

Liberty cap

Psilocybe semilanceata (Fr.) P. Kumm.

- ⊙ **Cap:** *0.75 inches (2 cm) wide, often shaped like an "elf hat," ochre-cream to ochre-yellow, with greenish or bluish tones near the edge.*
- ⊙ **Gills:** *adnate, dark brown-gray with white edging; spore print dark brown-violet.*
- ⊙ **Stem:** *cream or brownish with a greenish base.*
- ⊙ **Flesh:** *cream to brownish at the bottom of the stem; faint odor and sweet flavor.*
- ⊙ **Habitat:** *meadow and prairie grass, especially in mountainous regions.*
- ⊙ **When to observe:** *Especially in autumn.*

The liberty cap, reported from North America and the UK, is not common. Consuming it is extremely dangerous, and it is worth mentioning that laws in many countries forbid harvesting, transporting, or eating this mushroom, which was banned in the US in the early 2000s. Ingesting it produces several effects: euphoria, hallucinations, anxiety, and panic, sometimes with a transition to violent actions. In the event of an overdose, these symptoms may be accompanied by convulsions and even comas that can lead to death.

Did you know?

Without a doubt, the mycologist who has studied hallucinogenic mushrooms most extensively is Roger Heim from France. His work was primarily focused on the "sacred mushrooms" used by certain Mexican populations during traditional initiation ceremonies, described by missionaries as early as the sixteenth century. In his *Historia de las Indias de Nueva España* ("History of the Indies of New Spain"), Dominican friar Diego Durán discusses the holy ceremonies of Montezuma II, the ninth king of Tenochtitlan. He writes that "when the sacrifice was finished [...] they all went to eat raw mushrooms, food that made them lose all reason [...] Many killed themselves, and because of the power of these mushrooms they had visions and the future was revealed to them . . ."

Yellow
knight

Tricholoma auratum (equestre) (L. : Fr.) Gillet

- ◉ **Cap:** *reaching 5 inches (12 cm) in width, all yellow or greenish-yellow at first, then changes to red when exposed to light, often with fine and small brown scales.*
- ◉ **Gills:** *emarginate, crowded, a beautiful bright yellow; spore print white.*
- ◉ **Stem:** *fibrous, yellow, smooth.*
- ◉ **Flesh:** *white; strong flavor and odor of fresh flour.*
- ◉ **Habitat:** *under pine trees, often in the sand.*
- ◉ **When to observe:** *Autumn.*

The yellow knight (now considered a synonym for *T. equestre*) is reported from North America and the UK, and was widely considered an excellent edible until a few years ago. Unfortunately, when consumed in large quantities or over several consecutive meals, it can lead to a serious poisoning called rhabdomyolysis syndrome. Symptoms include extreme fatigue, high levels of perspiration, and intense muscular pain and rigidity caused by the progressive destruction of certain muscles. These symptoms can cause death. Now this mushroom is widely collected in some countries, while others consider it to be poisonous.

Warning

The story of the yellow knight should convince mushroom lovers once and for all that mushrooms should always be consumed in small quantities and never over several consecutive meals. This golden rule must be known and respected, even and especially with mushrooms that are considered edible. Several scientific articles published since 2005 report that a number of edible mushrooms, including the chanterelle (p. 113) and the king bolete (p. 129), are capable of causing rhabdomyolysis if consumed in large quantities.

False
morel

Gyromitra esculenta (Pers. : Fr.) Fr.

- ⊙ **Cap:** *shaped like a brain, 2.5 inches (6 cm) wide, red-brown to reddish russet color; spore print white to ochre.*
- ⊙ **Stem:** *chambered, whitish-brown, furrowed.*
- ⊙ **Flesh:** *fairly brittle; strong, pleasant odor and sweet flavor.*
- ⊙ **Habitat:** *in conifer forests, in soil rich in wood debris.*
- ⊙ **When to observe:** *Only in the spring, except at high elevations or cold regions.*

The false morel, reported from across North America and the UK, is not a common mushroom but grows abundantly in certain areas, particularly in the mountains in warmer climates. It is an attractive and appetizing mushroom that was once (and still is) widely consumed, but it has caused a few very serious poisonings and is now considered deadly. It appears that these accidents were caused by mushrooms that had not been properly cooked, and as a result the toxic substance, monomethylhydrazine, was not destroyed as it usually is by heat or dehydration. Nevertheless, it is best to be cautious and avoid consuming this mushroom altogether, especially since it has also been shown to be carcinogenic.

Warning

The false morel is confused with morels on a regular basis because they grow at the same time of year. Distinguishing it from its edible cousins, however, is quite easy: morels have a cap riddled with holes called alveoli, while the false morel's cap is made up of twisting grooves resembling those found in the brain. The stem of a morel is hollow, while that of the false morel is marbled with pits and flesh.

Giant
false morel

Gyromitra gigas (Krombh.) Cooke

- ⊙ **Cap:** *in the shape of a brain, often very irregular, reaches 5 inches (12 cm) wide, brown, cinnamon to ochre; spore print whitish.*
- ⊙ **Stem:** *chambered, whitish, stubby, heavily furrowed, often very irregular and curved.*
- ⊙ **Flesh:** *fairly brittle; strong, pleasant odor and sweet flavor.*
- ⊙ **Habitat:** *in forests with sandy or calcareous soil.*
- ⊙ **When to observe:** *Only in the spring.*

Very close to the false morel (previous page) and just as toxic, the giant false morel is fond of forests with sandy calcareous soil. Not reported from the UK, it is far less commonly reported in North America than *G. esculenta*. It is also larger than the false morel and its deformed cap and stem often take on strange shapes. Its cap is lighter than the false morel's. The pouched false morel (*G. infula*), reported from North America and the UK, is similar but its cap is made up of membranous lobes that look more like a turban than the grooves in a brain. It grows in autumn.

Did you know?

To distinguish the false morel from the giant false morel, mycologists often use a microscope. The spores of the two species are very different and easy to tell apart: the false morel's spores are elliptical and smooth, while those of the giant false morel possess a delicate network pattern and are elongated at each end by a characteristic blunt appendage.

- **Mushroom** usually in the shape of a small black "banana" that takes the place of a kernel in certain members of the Poaceae or Gramineae plant families by acting as a parasite.
- **Habitat:** in ears of wheat, rye, and numerous other wild species.
- **When to observe:** Late spring and in the summer.

Ergot

Claviceps purpurea (Fr.) Tulasne

This small mushroom, reported from North America, the UK, and Central Europe, appears fairly harmless, but is actually quite sinister. When it parasitizes cultivated grains, it poisons the flour that is made from them. The bread, as a result, becomes highly toxic and causes a condition known as "Saint Anthony's Fire." Individuals who have consumed poisonous bread usually experience convulsions, vomiting, and hallucinations, and the small blood vessels in their extremities become narrow (vasoconstriction). The resulting necrosis in the fingers and toes causes them to dry up and fall off "like charred wood," as one physician observed. Fortunately, today's farming methods have all but eliminated ergot from grains in most developed countries.

Did you know?

The ergot that appears in ears of *Gramineae* plants is only the first step in the *Claviceps* life cycle: formed in the spring, this "sclerotium" will fall to the ground with the kernels. The following year, after being soaked by the spring rains, long stems with small spherical heads will develop on its surface and produce spores that will eventually germinate in order to produce a new sclerotium.

▲ Devil's fingers

Bizarre Mushrooms

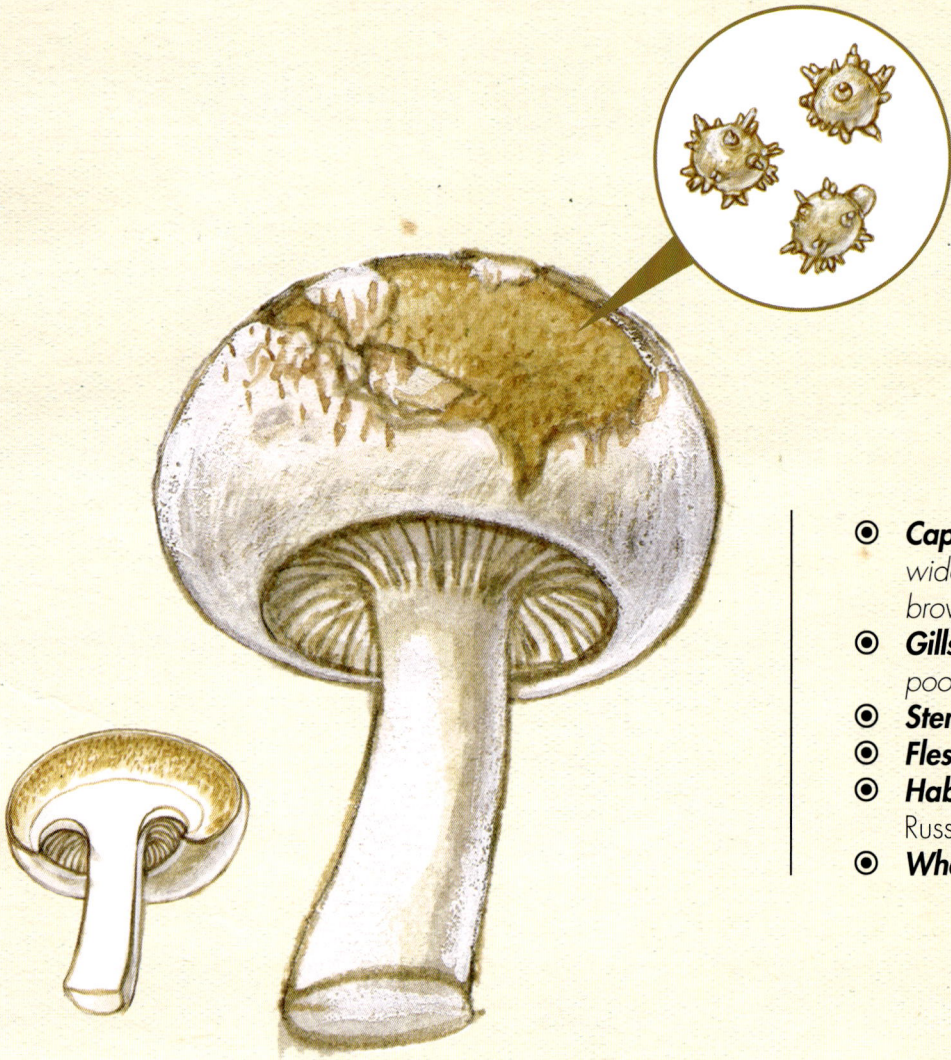

- ◉ **Cap:** *reaches just over 1 inch (3 cm) wide, covered entirely by an abundant brownish dust.*
- ◉ **Gills:** *adnate, beige, spaced and often poorly formed.*
- ◉ **Stem:** *white.*
- ◉ **Flesh:** *thin; flour-like flavor and odor.*
- ◉ **Habitat:** *usually on old decomposing* Russula *or* Lactarius *specimens.*
- ◉ **When to observe:** *Summer into autumn.*

Powdery piggyback

Asterophora lycoperdoides (Bull.) Ditmar

This small mushroom is commonly reported from North America and the UK, but nevertheless requires a bit of searching when the colors on the forest floor have been dulled by the first frosts at the end of autumn. It grows on old decomposing *Russula* specimens, and atop these cadavers it forms small brownish patches that you may overlook at first. Up close, it can easily be recognized by its powdery cap, its poorly formed gills, and the flour-like odor of its flesh. The silky piggyback (*A. parasitica*), which also grows on

▲ *Silky piggyback*

old *Russula*, has a silky cap rather than a powdery one and its gills are better formed.

Did you know?

The lyrical French name for this little mushroom ("Nyctalis porteur-d'étoiles," or "Nyctalis star-carrier") may sound a little nonsensical to some people, but it comes from the fact that when the powder on the cap is looked at under a microscope, it is made up of thousands of thick-walled brown cells (chlamydospores) that resemble stars. These cells function like true spores and are capable of germinating to form a complete mushroom.

Umbrella
polypore

Dendropolyporus (Polyporus) umbellatus (Pers.) Jülich

- ◉ **Mushroom can exceed 20 inches (50 cm) wide**, made up of a multitude of intertwining caps, each one capable of reaching 2 inches (5 cm) in diameter, sunken in the center, fibrillose, brownish-beige to ochre-cream.
- ◉ **Tubes:** thin, cream, decurrent; spore print white.
- ◉ **Stem:** fleshy, white, branching out beneath each small cap and inserting itself in the center.
- ◉ **Flesh:** brittle, white; sweet and pleasant odor and flavor.
- ◉ **Habitat:** on the floor of deciduous forests, in relationship with dead wood, often near stumps.
- ◉ **When to observe:** Summer to early autumn.

Few mushrooms, and even fewer polypores, reach the incredible size of the umbrella polypore, reported from North America and the UK. The giant polypore (p. 267) is also found on old stumps but only forms a few thick caps that sometimes grow to an imposing size (as much as three feet wide) and has pores that turn black when bruised or aging. Like the umbrella polypore, the hen-of-the-woods (see next page) is made up of many caps, but each one's stem is attached to the edge instead of its center, similar to what is observed among *Pleurotus* species.

Did you know?

The umbrella polypore is one only a few mushrooms that spend the harsh season underground as a mass of mycelium that looks like a leathery black tuber and is called a sclerotium.

Hen-of-the-woods

Grifola frondosa (Dicks.) S. F. Gray

- ⊙ *Mushroom can reach 20 inches (50 cm) and even 3 feet in diameter*, made up of a multitude of intertwining caps, each one reaching 2.75 inches (7 cm) wide and shaped like a spatula, gray-brown.
- ⊙ *Tubes:* rather thin, cream, decurrent; spore print white.
- ⊙ *Stem:* fleshy, white, branching out beneath each small cap and attaching itself laterally to each one (like Pleurotus).
- ⊙ *Flesh:* brittle, white; sweet and pleasant odor and flavor.
- ⊙ *Habitat:* at the foot of trees, especially oaks.
- ⊙ *When to observe:* Late summer into autumn.

This large polypore, reported from the UK and North America, is not very common but can grow for many years in the same spot. It is easily recognized and only likely to be confused with the umbrella polypore (see previous page), whose small caps have a different appearance and whose stem is inserted centrally beneath each cap, not on the side, or the giant polypore (p. 267) which has flesh that blackens. The brittle flesh of the hen-of-the-woods has a pleasant enough taste, but it remains a mushroom that is very rarely consumed in Europe. Eaten raw, it is toxic, causing gastric upset.

Did you know?

The hen-of-the-woods is known in Japan as the *maitake* and is one of the most commonly used mushrooms in traditional cuisine. It has been proven to contain substances that stimulate the immune system and favor the recovery of patients with certain cancers. It also possesses hypoglycemic effects used to treat diabetes.

Dyer's
polypore

Phaeolus schweinitzii (Fr.) Pat.

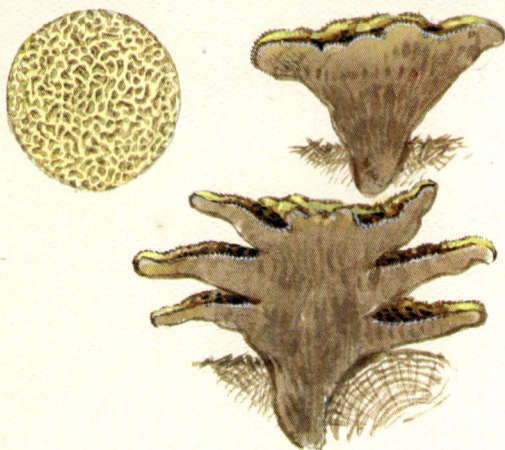

- ◉ **Cap:** *can reach 12 inches (30 cm) wide, soft and very velvety, dark rusty brown, mahogany brown, bright yellow border.*
- ◉ **Pores:** *irregular, "mustard" yellow-green then blackish-brown, turn brown when bruised; spore print yellowish-white.*
- ◉ **Stem:** *generally robust, rusty brown and felty.*
- ◉ **Habitat:** *at the base of conifers (pines, spruces, etc.) but also on their stumps, trunks, and roots.*
- ◉ **When to observe:** *Summer and autumn.*

The dyer's polypore, reported from North America and the UK, is very easy to recognize: the yellow hues along the edge of its cap and on its pores, its velvety cap, its soft flesh, and its habitat are all characteristic features.

It is so saturated with water when living that once it is dehydrated it becomes remarkably light. It can be a formidable parasite on weakened trees and hastens their decline. A few other polypores also have a felty cap, but they do not have stems and are much more leathery.

Did you know?

There are many mushrooms that can be used to dye wool and fabric. The dyer's polypore is especially appreciated for this in northern European countries. Depending on the part of the mushroom, the mordant, and the amount used, the colors obtained can vary from yellow to green or rusty brown. The colors are bright and deep and stand up well to washing and sunlight.

Dryad's
saddle

Polyporus squamosus (Huds.) Fr.

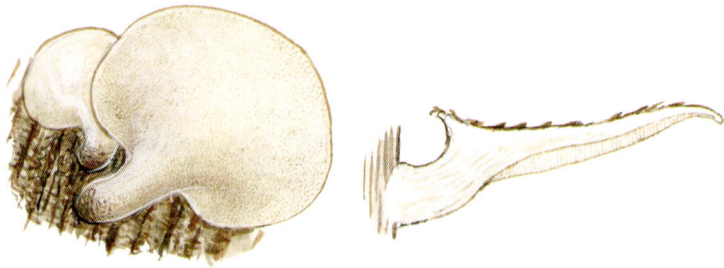

- ◉ **Cap:** *fan-shaped, can reach 20–32 inches (50–80 cm) wide, fairly thick but flexible, covered in brown scales on a cream to ochre-cream background.*
- ◉ **Tubes:** *rather broad and more or less polygon-shaped, whitish or ochre-cream; spore print white to pale yellow.*
- ◉ **Stem:** *distinct, solid, with a blackish base as it ages, not often centered in the cap.*
- ◉ **Flesh:** *fairly leathery, cream; sweet flavor with a slightly flour-like or mushroom odor.*
- ◉ **Habitat:** *on wood from living or dead deciduous trees.*
- ◉ **When to observe:** *Spring, summer, and autumn.*

The dryad's saddle, reported from North America and the UK, is not common, but is easily recognized by its large size and its cap covered in brown scales. It is a parasite on weakened trees and a major decomposer of dead wood. *Polyporus forquignonii* (*tuberaster*), a miniature version of the dryad's saddle, is much more common (reported from the UK and North America) and grows on the ground, on trunks, logs, or small branches on the forest floor. Its cap is covered with scales that are rarely more colorful than the background they rest upon, making them less visible than the scales on the dryad's saddle, and its stem inserts itself more or less centrally on the cap.

Did you know?

There are several other species genus *Polyporus*, and all of them are characterized by a well-formed stem that is usually inserted in the center of the cap. The fringed polypore (*P. ciliatus*) (synonym: *Lentinus substrictus*), reported from the UK, and North America, is a small and elegant mushroom with the edges of its cap covered in fine hairs arranged in a regular pattern and very thin pores that are almost invisible to the naked eye.

▲ *Polyporus forquignonii*

▲ *Fringed polypore*

- ⊙ **Cap:** 6 inches (15 cm) wide, covered in large wooly pyramidal scales, brown-gray then blackish on a pale gray background.
- ⊙ **Tubes:** rather wide, pale gray then gray-brown, reddens when bruised; spore print black.
- ⊙ **Stem:** the same color as the cap, fluffy and fleecy with a distinct ring, reddens when bruised.
- ⊙ **Flesh:** whitish to grayish, reddens when cut; sweet flavor and faint odor.
- ⊙ **Habitat:** under deciduous trees (especially beech), sometimes under conifers, especially in mountainous areas in warmer regions.
- ⊙ **When to observe:** Summer to autumn.

Old-man-of-the-woods

Strobilomyces strobilaceus (Scop. : Fr.) P. Karst.

The enormous bolete group

The boletes are one of the mushroom groups that the general public is most familiar with, but they are also one of the most delicate groups where mycological study is concerned. More than thirty different bolete genera exist around the world, and some of them can only be distinguished by subtle microscopic details. What is more, precision morphology techniques using DNA fragment comparison have regularly demonstrated that a number of species in tropical regions in particular are still unknown to scientists and have yet to be described.

The old-man-of-the-woods, reported from the UK and North America, is not common but is very easy to recognize. Its colors, the "tattered" appearance given by its wooly scales, its gray pores, its ring, and its reddening flesh make it one of the most generous boletes in terms of distinctive features. It is the only European *Strobilomyces*; the other species in this genus are scattered throughout North America, the tropics, and equatorial regions. In eastern North America, where both are reported, it is very difficult to distinguish *S. strobilaceaus* from *Strobilomyces confusus*. And in eastern North America, some mycologists believe that *S. floccopus* is a separate, similar species. One day the argument may be settled!

Edible . . . perhaps!

The old-man-of-the-woods is sometimes considered edible, but its "dirty" appearance is hardly appealing. It is also sufficiently rare that harvesters are encouraged to protect it rather than pick it.

Did you know?

Spore shape and ornamentation are critically important features for bolete classification. While the large majority have smooth, elongated, elliptical spores, species from genus *Boletellus*, for example, display elongated spores with longitudinal ridges. *Strobilomyces* species, on the other hand, form spherical spores with large spines or crests that form a network pattern.

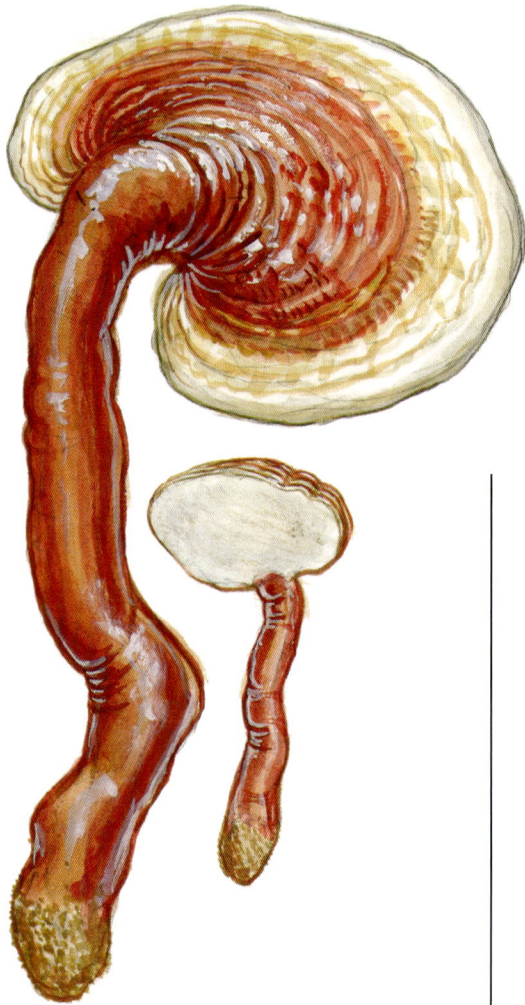

Lacquered bracket

Ganoderma lucidum (Curtis) P. Karst.

- ◉ **Cap:** *reaches 4 inches (10 cm) in width, brown-red to brown-yellow along the edge, lacquered appearance, irregular concentric circles that form bulges.*
- ◉ **Tubes:** *rather thin, cream or tobacco brown; spore print brown.*
- ◉ **Stem:** *more or less developed but always present, inserted at the edge of the cap, the same color as the cap, roughly cylindrical but often very irregular and nodulated.*
- ◉ **Flesh:** *light brown, cork-like consistency, tough; flavor and odor not distinctive.*
- ◉ **Habitat:** *on dead deciduous wood, rarely on living trees.*
- ◉ **When to observe:** *Begins fruiting in early spring, harvest at maturity late summer through autumn.*

The lacquered bracket, reported from North America and the UK, seems like it should be easy to recognize with its unique coloring and appearance, but several closely related species bear a striking resemblance to it and telling them apart requires experience. *Ganoderma tsugae*, reported from northeast North America and found on Hemlock; *G. lucidum*, reported from North America and the UK and found on deciduous trees; *G. curtisii*, reported from northeastern and southwestern North America and also found on deciduous trees; *G. sessiliforme*, rarely reported (and only from North America) which has no stem; *G. oregonense*, reported from western North America and found on conifers; and *G. resinaceum*, reported from eastern and central North America as well as the UK, which does not usually have a stem and has a reddish crust on its cap that melts when exposed to flame, are all almost perfect lookalikes. *G. valesiacum* has a distinct stem but is typically a lighter red and only grows at the base of old larch trees (genus *Larix*) in mountainous areas in south central Europe.

Did you know?

The lacquered bracket, like some of its peers, has been used for centuries in traditional Asian medicine. Recent studies have shown that it produces a number of active substances, several of which are beneficial in the treatment of certain cancers. Clinical trials in humans are only beginning, of course, but it is possible that these mushrooms will one day be included in the treatment protocols for these diseases.

Red-belted
bracket

Fomitopsis pinicola (Swartz) P. Karst.

- ⊙ **Cap shaped like a hoof**, *can reach 10 inches (25 cm) in width, hard, smooth, marked with concentric circles, dark brown-gray near its insertion onto the wood but tinted orange to orangey-red along the edge in the growth zone.*
- ⊙ **Pores:** *very thin and crowded, yellowish to cream; spore print white.*
- ⊙ **Stem:** *absent.*
- ⊙ **Flesh:** *cream to ochre, extremely leathery; flavor not distinctive, but a powerful odor that is difficult to describe, both resinous and sweetish.*
- ⊙ **Habitat:** *on dead wood from deciduous and coniferous trees, rarely found on living trees.*
- ⊙ **When to observe:** *All year long, but especially in autumn.*

The red-belted bracket, reported from North America and the UK, is fairly common and can easily be identified by the various shades of red and orange on its cap. It also gives off a very unique odor that is hard to describe. When it ages and these radiant colors start to disappear, it may be confused with the hoof fungus (p. 265), though the latter has much paler cream to ochre-colored flesh and pores. Even rarer is *Fomitopsis rosea* (*Rhodofomes roseus*), reported from the same regions (though less commonly), which looks a great deal like the red-belted bracket though its pores are a soft pink.

▲ *Fomitopsis rosea*

Did you know?

There are many polypores—from the Greek *poly*, "several," and *poros*, "pore, passage"—and they are not always easy to identify. Mycologists traditionally divide them into two major groups: polypores with a cap and polypores said to be "resupinate" that do not form caps. The second group is large and highly diversified, forming crusts pierced with many pores on the underside of trunks or fallen branches lying on the ground.

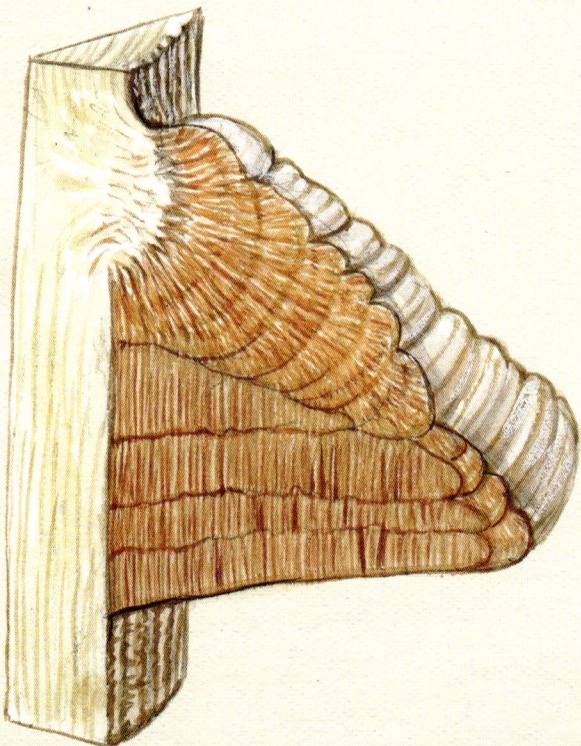

- ⊙ **Cap shaped like a hoof**, can reach 10 inches (25 cm) in width, hard, smooth, marked with concentric circles, dark gray-brown to dull ochre-brown.
- ⊙ **Pores:** very thin and crowded, brown; spore print white to pale yellow.
- ⊙ **Stem:** absent.
- ⊙ **Flesh:** brown, very leathery, cork-like consistency, with a tender "core" where the fungus inserts onto the wood; flavor and odor not distinctive.
- ⊙ **Habitat:** on living or dead deciduous trees, rarely on conifers.
- ⊙ **When to observe:** All year long.

Hoof fungus

Fomes fomentarius (L. : Fr.) Fr.

The hoof fungus, reported from North America and the UK, is probably the most common and most widespread of the large polypores, but is nevertheless rare in some areas, particularly in the mountains. It is easily recognized by its cap marked with concentric circles and its brown flesh, which has a tough texture like cork. It sometimes appears in a much lighter form known as variety *inzengae* that is a brownish ochre cream.

Ötzi, a mycologist from another age

In 1991, the mummified and frozen body of a shepherd was discovered inside a mountain glacier. The corpse was a little over 4,500 years old. Perfectly preserved by the cold, Ötzi—as he was named—still had all of the equipment a proper Neolithic nomadic shepherd would have carried: a bow, arrows, and a knife with a flint blade, along with a small sack containing everything needed to make a fire, including flint and amadou, or tinder fungus. This flammable substance is made by reducing the flesh of the hoof fungus to a powder. The Latin name for the hoof fungus, Fomes, comes directly from the Latin and means "flammable substance."

Warning

There are several other large hoof-shaped polypores found frequently on living and dead trees. The red-belted bracket (p. 263) strongly resembles the hoof fungus but has orangey to reddish-orange colors along the edge of the cap (its growth zone) and its pores are cream to yellowish. The artist's bracket (*Ganoderma lipsiense*) (*Polyporus lipsiensis*), not reported from North America or the UK, has a cinnamon to ochre-brown cap often covered in a rusty brown dusting and white spores that are frequently deformed by small pustular warts. It is interesting to note that this "dust" is formed by the mushroom's spores, which are deposited *en masse* on the cap after being liberated by the tubes. In North America and the UK, the artist's conk is *Ganoderma applanatum*.

Did you know?

Many large polypores like the hoof fungus develop just as well as parasites on living trees as they do as saprotrophs on dead ones (from the Greek *sapros*, "rotten, putrid," and *troph* "food or nourishment," used to characterize organisms that develop on decomposing organic matter). In truth, however, most of them are unable to grow on trees that are in perfect health and only establish themselves on weakened or wounded specimens. They are called "weak parasites" for this reason and it is important, especially in more urban areas, not to prune trees too severely, because every wound is an open door for mushrooms like these to enter.

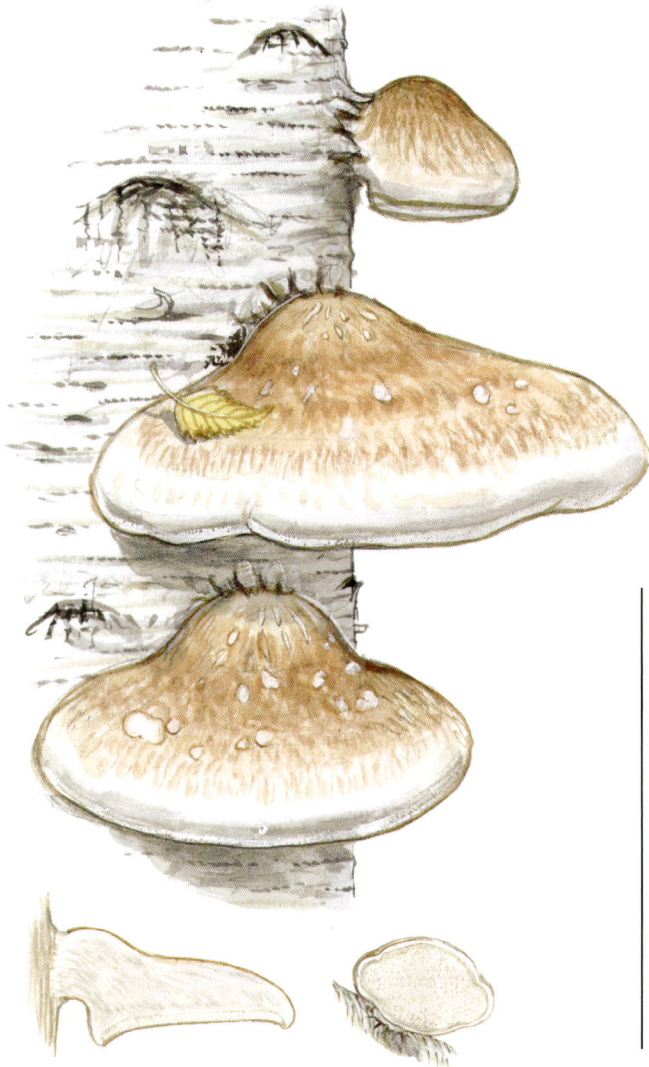

Birch
polypore

Piptoporus betulinus (Bull. : Fr.) P. Karst.
(*Fomitopsis betulina*)

- ◉ **Cap in shape of a hoof**, *can reach 8–12 inches (20–30 cm) wide, very rounded edge, not very hard, smooth or somewhat felty, whitish or ochrebrown.*
- ◉ **Tubes:** *very thin, white or whitish; spore print white.*
- ◉ **Stem:** *reduced but still distinct, formed where the cap becomes slimmer at the point of insertion onto the wood.*
- ◉ **Flesh:** *rather soft, cream; sweet flavor and woody odor.*
- ◉ **Habitat:** *only on dead birch wood.*
- ◉ **When to observe:** *Summer and autumn.*

This polypore, reported from North America and the UK, is a traditional example of some mushrooms' extreme specificity: it only grows on birch, and this preference is what has given it its name.

Unlike many polypores growing on dead wood, it detaches easily from its substrate because the area connected to the wood is reduced to a kind of small stem. At one time, specimens in perfect condition would be left out to dry (this mushroom's soft flesh can rot rather quickly) and then used to sharpen razor blades with their leather-like surface.

Did you know?

The birch polypore is an annual mushroom and grows and dies every season, unlike other polypores that continue growing over several years. It has to be tough in order to remain in place during bad weather, and specimens "mummified" by the frost are often found in winter. A mushroom that causes the birch polypore to decay more quickly is sometimes found on its inferior face. *Hypocrea pulvinata* (*Trichoderma pulvinatum*), more uncommonly reported than its host, forms small powdery ochre-yellow cushions that are sometimes hard with black spots.

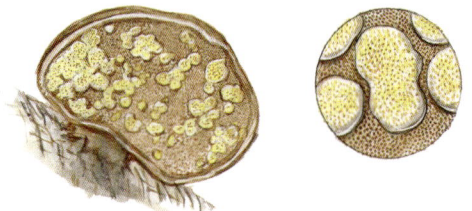

Giant
polypore

Meripilus giganteus (Pers.) P. Karst.

- ⊙ **Caps that are fleshy and large,** layered inside one another in fan shapes reaching 32–36 inches (80–100 cm) wide, marked with concentric circles, brown-beige to ochre-brown, turn black with age or when bruised.
- ⊙ **Pores:** thin, whitish or cream, turn black; spore print white.
- ⊙ **Stem:** reduced to a massive central trunk.
- ⊙ **Flesh:** rather soft and easy to break, chalky consistency as it dries; somewhat bitter flavor and pleasant mushroom odor.
- ⊙ **Habitat:** on deciduous trees, most of the time on old stumps.
- ⊙ **When to observe:** Summer into autumn.

The giant polypore, reported from North America and the UK, is not very common but certainly does not go unnoticed. Its caps tend to grow on logs of substantial proportions—given the size that this organism is likely to attain—and usually feed on the dead wood of the substrate. It may be confused with chicken mushroom (p. 269) specimens that have lost their beautiful colors but can be distinguished by its blackening. A very similar species, *M. sumstinei*, is reported from the same range.

Did you know?

Mushrooms that feed on dead wood produce substances called enzymes that break down the long molecules of cellulose and lignin found in wood fibers. Not all of them, however, produce the same enzymes or the same quantities: certain polypores only break down cellulose, while others focus exclusively on lignin. As a result, the rotting wood the mushroom feeds on may appear one of two ways: in the case of brown rot, the wood is fragmented into small cubes after the mushroom has attacked the cellulose, and in white rot, the wood takes on a fibrous look as a result of the polypore's destruction of the lignin.

- ⊙ **Mushroom usually made up of several fan-shaped caps**, able to reach 3 feet (1 m) wide, bright orangey-yellow to yellow along the edge, undulating, paling in color as it ages and dries.
- ⊙ **Pores:** thin and crowded, sulphur yellow then paler; spore print white to cream.
- ⊙ **Stem:** fleshy, central, short.
- ⊙ **Flesh:** pale yellow, brittle especially when dry; pleasant flavor and odor.
- ⊙ **Habitat:** at the base of living and dead trees, particularly oaks.
- ⊙ **When to observe:** Late spring, summer and autumn.

Chicken Mushroom

Laetiporus sulphureus (Bull.) Murrill

If there is one polypore that is easy to recognize, reported from North America and the UK, it is the chicken mushroom. Its tremendous size and vibrant colors are unique characteristics that make it difficult to mistake for anything else. Still, as it ages, those beautiful colors tend to disappear and it then begins to look like the giant polypore (p. 267). The giant polypore's caps, however, are not undulated, less delicate, and turn black.

Warning

The chicken mushroom is found just about everywhere in the world and is traditionally considered edible. Its English common name comes from the fact that its flesh has the consistency of a chicken breast. It has nevertheless been blamed for mild cases of poisoning, so it is best to be cautious, especially because this name is sometimes misapplied to several different species that are hard to tell apart: *L. huroniensis*, reported from eastern and central North America, is found on conifer and considered poisonous; *L. cincinnatus*, reported from eastern and central North America, has a pink cap, cream pores, and is considered a synonym of *L. sulphureus*; *L. gilbertsonii*, reported from western North America, can be found on Eucalyptus; *L. conifericola*, also reported from western North America, can be found on conifer.

Did you know?

Because mushrooms like the chicken mushroom are found in nearly every country in the world and in both hemispheres, people often assume that they are extremely flexible when it comes to their habitat. In recent years, though, more and more comparative DNA studies have shown that contrary to this belief, the specimens on different continents are genetically quite different and many of them represent distinct species. As we have seen, morphological characteristics in mushrooms can very often be deceiving.

House mushrooms

No less than 100 mushrooms are capable of growing in our homes, especially if a water leak provides enough humidity for the spores already present in the atmosphere to germinate. Dry rot (Serpula lacrymans), reported from North America and the UK, is probably the most fearsome, and can destroy wooden framing and entire wood floors within weeks, leading to costly repairs. The chicken mushroom, though not often found in homes, can appear spontaneously on handcrafted wood pieces if they are damp enough.

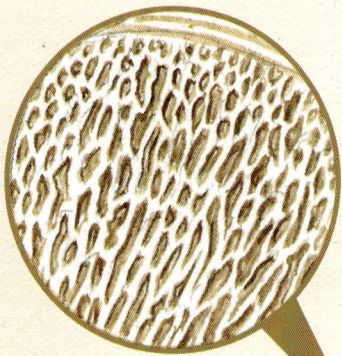

- ⊙ **Cap shaped like a semicircle**, can reach 8 inches (20 cm) in diameter, white or whitish, felty, often turned green by microscopic algae, irregularly bumpy especially near the insertion onto the wood.
- ⊙ **Tubes:** white and fairly large, irregular and elongated, separated by thick walls; spore print white.
- ⊙ **Stem:** absent.
- ⊙ **Flesh:** very leathery, white; flavor and odor not distinctive.
- ⊙ **Habitat:** on various dead deciduous trees, often on hornbeams and beech.
- ⊙ **When to observe:** All year.

Lumpy
bracket

Trametes gibbosa (Pers. : Fr.) Fr.

The lumpy bracket, reported from the UK and North America, is a common mushroom whose pale semicircles are often found adorning younger tree stumps. The surface of its cap is velvety and home to microscopic aerial algae that turn it green as they develop large colonies. The pores are not very large but are distinctly elongated and separated by thick partitions, a helpful indicator for identifying this polypore. It might be confused with *Trametes pubescens*, which is also pale, but the latter is much smaller, its caps are marked with radially arranged fibers, and its rounded pores are smaller, or *Dadaelia quercina*, which has very large, gill like pores, and very thick partitions. Both are reported from the same regions as *T. gibbosa*.

Did you know?

When out looking for mushrooms, specialists are often faced with

an age-old and dreaded question that encapsulates the priorities of most mushroom lovers all too well: "Is this edible?" Even polypores, which are hard as wood for the most part, cannot escape this fateful interrogation, and some irritated mycologists are quick to reply with the following: "Of course, all polypores are edible. . . but only when cooked with pebbles. When the pebbles are cooked, then you can eat the polypores!"

▲ *Trametes pubescens*

Turkeytail

Trametes versicolor (L. : Fr.) Lloyd

- ◉ **Caps in fan shape**, *very leathery but thin and supple, tend to grow in colonies and form wreaths that are sometimes quite large; highly variable in color, going from brownish-cream to blackish through every shade of brown and grayish-blue, slightly velvety with concentric circles.*
- ◉ **Tubes:** *thin, white; spore print white.*
- ◉ **Stem:** *absent.*
- ◉ **Flesh:** *thin, very leathery, white; flavor and odor not distinctive.*
- ◉ **Habitat:** *on various deciduous trees, but also sometimes on conifers.*
- ◉ **When to observe:** *All year.*

The adjective "versicolor" means "presents varied colors," and perfectly describes this extremely common *Trametes*, reported from North America and the UK. A challenging mushroom for the uninitiated, the turkeytail cannot be identified by its color because it is so variable and instead must be recognized by its thin and flexible caps, with zonate colors and textures, from shiny to velvety, and thin white pores. The smoky bracket (*Bjerkandera adusta*), reported from the same regions, resembles the turkeytail, but is less common, has thicker flesh, and has unique mouse gray gills.

Did you know?

Nine species from genus *Trametes* exist in Europe (plus at least that many in North America, dependent upon which genus mycologists decide to put some of the species) and the most common of them is without a doubt the turkeytail. No matter what time of year it is, if you go out into the woods you are almost guaranteed to encounter one. It is also a rather cosmopolitan mushroom and grows in almost every part of the world.

▲ *Smoky bracket (Bjerkandera adusta)*

Giant
puffball

Calvatia gigantea (Batsch) Lloyd

- ◉ **Mushroom shaped like a large smooth ball**, *can reach 32 inches (80 cm) in diameter, fairly soft, white at first then ochre-brown as it ages.*
- ◉ **Flesh:** *soft, white with cheesy texture at first, then wet and yellowish, finally powdery and ochre-colored; unpleasant flavor and odor similar to certain rubbers; spores light brown.*
- ◉ **Habitat:** *fields and prairies, often in grazing areas.*
- ◉ **When to observe:** *Summer into autumn.*

It is impossible to confuse the giant puffball (reported from across North America and the UK) with any other mushroom: its shape, imposing size, and its habitat allow it to be seen from afar. Some people consider it edible and cook it in thick slices like steaks, taking care to only harvest young specimens with flesh that is still totally white. The giant puffball ripens quickly, transforming it into an unappetizing ochre-brown powder made up of billions of tiny spores (see adjacent). *Calvatia (Lycoperdon) utriformis* bears a slight resemblance, but has a large sterile base, and grows in the same regions and places as the giant puffball but is much smaller and covered in large flat and irregular warts. *Calvatia cyathiformis* and *C. craniiformis*, reported from North America, also have sterile bases, with a smoother skin than *C. utriformis*.

Did you know?

The giant puffball forms roughly 7 trillion spores. Placed end to end, the huge necklace they would form would measure 21,747 miles (35,000 km), or roughly four times the distance between Paris and Tokyo!

▲ *Calvatia utriformis*

▲ *Calvatia gigantea*

Wolf's milk
slime mold

Lycogala epidendrum (J. C. Buxb. ex L.) Fr.

- ⊙ **"Mushroom" shaped like small balls**, *soft and salmon pink at first, covered with tiny pustules and releasing a king of orangey-pink paste when crushed, then ochre to tobacco brown spores, turned powdery by dry weather or doughy (like modeling clay) by humid weather.*
- ⊙ **No stem.**
- ⊙ **Habitat:** *on dead wood or trunks lying on the ground.*
- ⊙ **When to observe:** *Just about all year long, except during freezing temperatures.*

This strange "mushroom" (reported from North America and the UK) is very common on dead wood but often goes unnoticed and is technically not a mushroom at all. It is actually a representative of a very unique group of organisms called the myxomycetes, commonly called "slime molds," that are no longer part of Kingdom Fungi as it is defined today. They are currently classified as protists, a large and very diverse group that is undergoing its own restructuring process. In spite of all of this, myxomycetes are still studied by mycologists, and a recent monograph of this group counted over 850 species in Europe.

Did you know?

The myxomycetes can easily be differentiated from true mushrooms by their biology. Instead of producing a filament (the hypha) during germination like the majority of organisms in Kingdom Fungi, their spores form a kind of amoeba that moves around on or through the substrate, feeding on organic debris and sometimes on other living organisms. At a certain point the amoebae from one organism come together and fuse to create larger organism called a plasmodium that continues to move until it is fixed in place and becomes the new organ of spore production.

Gem-studded
puffball

Lycoperdon perlatum Pers. : Pers.

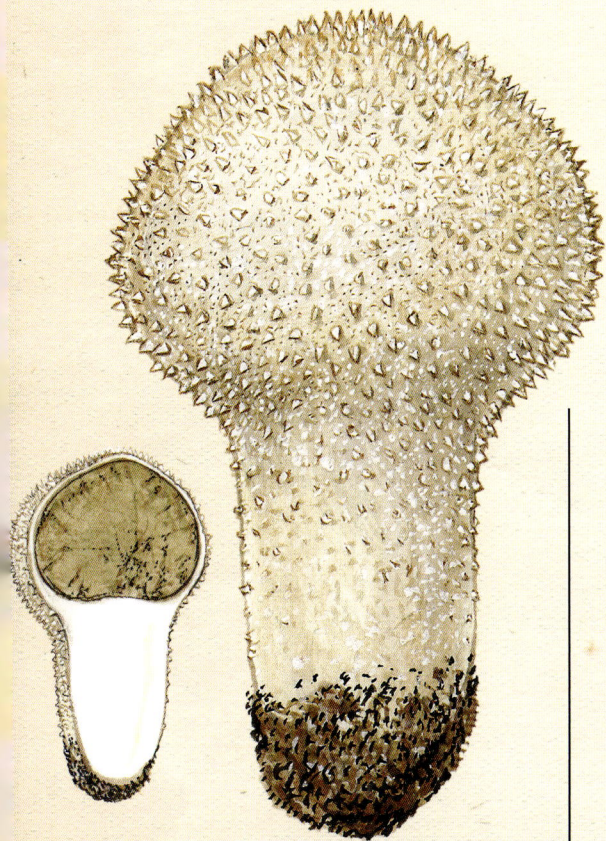

- ◉ **Mushroom in soft ball shape,** *0.40–2.5 inches (1–6 cm) tall, white, thorny with small labile pyramidal warts that leave behind a round scar surrounded by tiny granular warts; a hole (the ostiole) appears on the top of the ball in mature specimens.*
- ◉ **Stem:** *distinct, up to 1.20 inches (3 cm) tall, soft, whitish.*
- ◉ **Flesh:** *soft, white in the stem and the remainder of the mushroom, white at first in the upper portion (the gleba) then brown and powdery; sweet flavor and faint odor similar to rubber; spores yellow-brown to olive-brown.*
- ◉ **Habitat:** *on the ground in forests.*
- ◉ **When to observe:** *Summer and autumn.*

Mushroom lovers are generally familiar with puffballs, though telling them apart from each other is not always easy. The gem-studded puffball, reported from North America and the UK, is probably the most common and is recognizable by its fragile pyramidal warts, which fall off and leave behind a scar. The pear-shaped puffball (p. 276) has a similar appearance but is covered in fine brown labile granules that give it a smooth texture. It also grows in colonies and tufts on old stumps or heavily decomposed trunks.

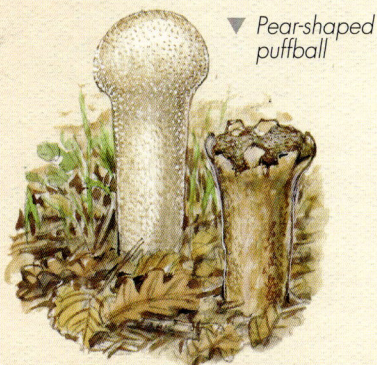

▼ Pear-shaped puffball

Did you know?

At least thirty puffballs exist in Europe and North America and according to mycologists they belong to at least four different genera. *Calvatia (Lycoperdon) excipuliformis*, reported from North America and from the UK, for example, looks like a larger version of the gem-studded puffball and can reach 8 inches (20 cm) tall.

Its very developed stem is almost completely rot-resistant and can remain in the same place on the forest floor long after the spore-producing upper portion has disappeared. Because of their rubbery consistency, these stems were once used as splints to protect bone fractures.

Pear-shaped puffball

Morganella (Lycoperdon) pyriforme (Fr.) Pat.

- ⊙ **Mushroom in soft ball shape,** *0.40–2 inches (1–5 cm) tall, cream to grayish-brown, covered in small warts or brown labile granules that appear delicately placed on a smooth surface; a hole (the ostiole) appears on the top of the ball in mature specimens.*
- ⊙ **Stem:** *distinct with a pear-shaped appearance, abundant white threads of mycelium, called rhizomorphs, at the base.*
- ⊙ **Flesh:** *soft, white in the stem and the remainder of the mushroom, white at first in the upper portion (the gleba) then olive-brown and powdery; sweet flavor and faint odor similar to rubber, spores olive-brown.*
- ⊙ **Habitat:** *in tufts or large colonies on dead wood, particularly stumps and massive trunks that are already decomposing.*
- ⊙ **When to observe:** *Summer and autumn.*

The pear-shaped puffball, reported from North America and the UK, can be recognized by its habitat: it is the only puffball in Europe that forms colonies on dead wood. There is another *Morganella*—a genus that includes lignicolous puffballs—named *M. (Lycoperdon) subincarnata* that grows abundantly in eastern North America, but it is so rare in Europe that few mycologists can claim to have seen it. The meadow puffball (*Vascellum pratense*), reported from North America and the UK, often grows in fields and lawns and bears a slight resemblance to the pear-shaped puffball with its smooth appearance, but its flatter shape and grassy habitat are enough to set the two apart.

Did you know?

In mushrooms with well-developed caps, gills, tubes, or teeth, the mature spores are violently ejected and dispersed over long distances (see p. 34). But in gasteromycetes (see p. 289), the spores are enclosed within a "stomach" that remains sealed, preventing the same kind of dispersal mechanism. Puffballs, on the other hand, put their trust in the elements and animals to disperse their spores: children who enjoy pinching ripened puffballs to see the "smoke" that escapes are unwittingly taking part in the scattering of millions of tiny spores.

Leopard
earthball

Scleroderma areolatum Ehrenb.

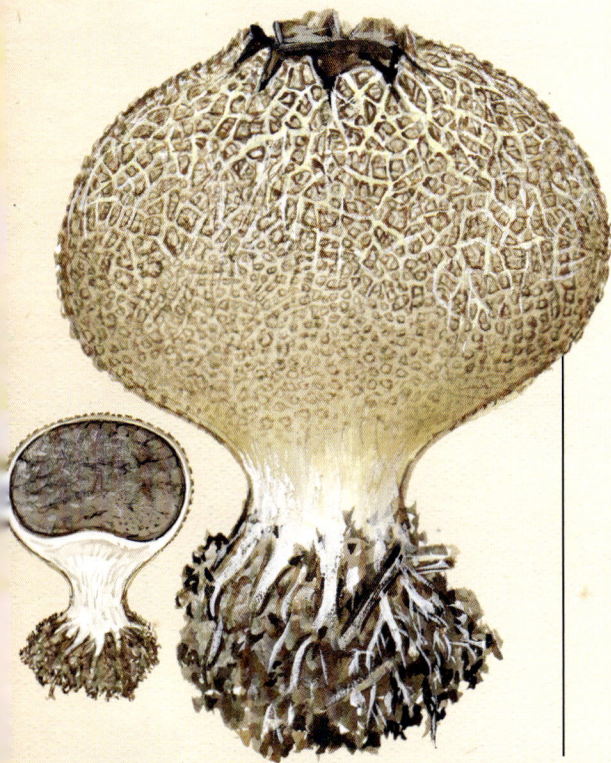

- ⊙ **Mushroom shaped like a ball**, 0.80–2.5 inches (2–6 cm) wide, light ochre-brown, cracked with small irregular patches showing the cream background.
- ⊙ **Stem:** fairly thick, whitish, made up of an accumulation of mycelium threads.
- ⊙ **Flesh:** gray or purple-black to gray-black with white veins at first, very firm, then dark brown and powdery; sweet flavor and faint rubbery odor.
- ⊙ **Habitat:** under deciduous and coniferous trees.
- ⊙ **When to observe:** Spring through autumn.

Three *Scleroderma* species are found frequently in and around mainland Europe: the leopard earthball, the scaly earthball (*S. verrucosum*), and the common earthball (*S. citrinum*); all also reported from North America and the UK. The scaly earthball resembles the leopard earthball and the two species are often confused, but the scaly earthball can be distinguished by its large irregular and flat warts that are very close together, making the lighter background behind them almost invisible. The common earthball, which is sometimes invaded by the mycelium of *Pseudoboletus parasiticus*, reported from North America and the UK, can be identified by its distinct yellow tones. There are at least four other earth ball species from the same regions, that are less commonly reported.

Did you know?

The name of genus *Scleroderma* comes from the Greek *scleros*, "hard," and *derma*, "skin." These mushrooms have a thick skin that sets them apart from puffballs (see previous) with their thin and fragile envelopes. The firm black flesh in young *Scleroderma* could be mistaken for the black truffle (p. 159), which has led some unscrupulous restaurant owners to use them in place of the "black diamond."

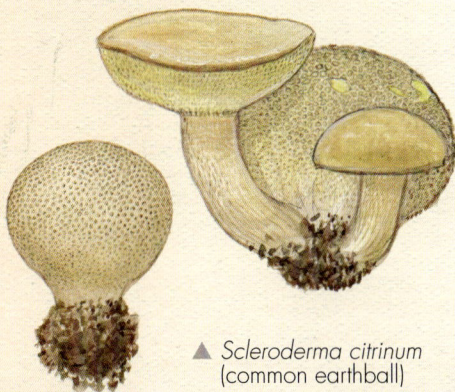

▲ *Scleroderma citrinum* (common earthball)

Cedar
cup

Geopora sumneriana (Curtis) P. Karst.

- ⦿ **Mushroom shaped like a sphere,** *buried almost entirely underground, piercing the surface of the ground by forming a star-shaped opening; velvety and brown to reddish-brown external face, smooth cream internal face.*
- ⦿ **Stem:** *absent.*
- ⦿ **Flesh:** *thin, brittle, whitish; flavor and odor not distinctive.*
- ⦿ **Habitat:** *especially under cedars, not as common under yew trees.*
- ⦿ **When to observe:** *Only in the spring.*

This small *Peziza*, reported from the UK but not from North America, would appear quite harmless if it did not cause gardeners such serious problems, particularly in urban parks. It can grow anywhere cedars are planted and sometimes forms colonies of several dozen individual specimens. Imagine for a moment if it was your job to maintain the grass in a park where several of these trees were growing and you should have some idea of the number of mushrooms that swell and perforate these lawns in the spring.

Did you know?

Several other *Geopora* exist, but fortunately they do not cause very many problems. *G. arenicola* (reported from North America and the UK), for instance, enjoys sandy soil and can often be found behind sand dunes. *Sarcosphaera coronaria* (reported from the UK and North America) is another springtime *Peziza* that is also partially subterranean but much larger: it can reach 6 inches (15 cm) in diameter and prefers very calcareous soil. This species can be recognized easily by its purple internal face.

▲ *G. arenicola*

▲ *S. coronaria*

▲ *G. summeriana*

Yellow
stagshorn

Calocera viscosa (Pers.) Fr.

- ⊙ **Mushroom** *up to 2 inches (5 cm) tall, shaped like multiple branches, bright orangey-yellow, very viscous in humid weather, then sticky and finally hard and horn-like as it dries.*
- ⊙ **Flesh:** *very leathery and elastic; faint flavor and odor.*
- ⊙ **Habitat:** *on dead deciduous and coniferous wood.*
- ⊙ **When to observe:** *Summer through early winter.*

The yellow stagshorn, reported from North America and the UK, is really quite easy to recognize. It grows on dead wood and its beautiful color makes it hard to miss. Harvesting it, though, is more complicated because its leathery yet extremely slippery texture renders it impossible to hold onto and it is anchored to its substrate as sturdily a mussel is to a rock. When dry it becomes almost unrecognizable—hardened and a dull orangey-brown color—but is still just as tough. Its close relative, the small stagshorn (*C. cornea*), reported from North America and the UK, forms colonies of small horns around 1 cm tall on dead and decaying wood. *C. furcata*, reported from the UK but rarely from North America, is found on conifer wood.

▼ *Calocera cornea*

Did you know?

While the yellow stagshorn's shape is reminiscent of certain *Clavaria* species (pp. 153–154, 223–225), it is part of another group entirely. Mycologists believe that *Clavaria* morphology has appeared independently several times over the course of evolution. There are numerous examples of these convergences in mushroom shape, and they present significant challenges for scientists trying to construct a classification system that accounts for species' true biological similarities.

Coral
tooth

Hericium clathroides (Pall.) Pers.

- ⦿ **Mushroom in coral shape**, *very elegant, with a branch-like structure reaching 16 inches (40 cm) in width, white or whitish, made up of many teeth, 1 cm in length, arranged in rows.*
- ⦿ **Flesh:** *brittle, white; sweet and pleasant odor and flavor; spore print white.*
- ⦿ **Habitat:** *on wide pieces of dead wood, usually beech wood.*
- ⦿ **When to observe:** *Summer and autumn.*

▲ *H. flagellum*

Like other *Hericium* species, the coral tooth is a mushroom typically found in preserved natural forests in Europe. Because such habitats in most European countries now only exist as relics in remote areas or in protected environmental reserves, it has become rare throughout most of Europe. It is therefore critical to protect its status. *H. clathroides* is not reported from the UK or North America. *Hericium flagellum*, not reported from North America or the UK, has a similar appearance but its teeth are arranged in all directions and it grows on fir and spruce trees. *H. coralloides*, reported from North America and the UK, has a very similar appearance to *H. clathroides*. *H. americanum*, reported from North America but not the UK, is very similar, can be found on wounds or dead wood of hardwoods, but has longer, more clustered teeth.

Did you know?

Specialists in forest ecology in Europe have established a list of species whose presence in a forest is an indication of its ecological character and the continuous presence of old trees. Monitoring "bioindicator" species like *Hericium* helps scientists assess the current conditions of preservation, environmental maturity, and habitat.

Bearded
tooth

Hericium erinaceus (Bull.) Pers.

- ◉ **Mushroom in coral shape**, *very elegant, not branched, in a regular ball shape that can reach 10 inches (25 cm) in diameter, made up solely of long parallel teeth that can be 2 inches (5 cm) long.*
- ◉ **Flesh:** *brittle, white; sweet and pleasant odor and flavor; spore print white.*
- ◉ **Habitat:** *on wide pieces of dead wood, especially beech wood.*
- ◉ **When to observe:** *Summer and autumn.*

Like all *Hericium*, the bearded tooth, reported from North America and the UK, is a rare mushroom. It is very easy to recognize because of its long parallel teeth. It could potentially be confused with young tiered tooth specimens (*H. cirrhatum*), reported from the UK and from North America, but the latter is usually made up of several layered caps and has shorter teeth up to 1.5 cm long. It is used medicinally in Asia to manage high blood sugar.

Did you know?

Hericium have been the topic of much discussion in Europe. The bearded tooth was one of 33 mushroom species put forward at the Bern Convention on the Conservation of European Wildlife and Natural Habitats and is on the protected species list in thirteen European countries. It is also a deciding factor in the classification of its habitats as Natural Areas of Ecological Fauna and Flora Interest (ZNIEFF).

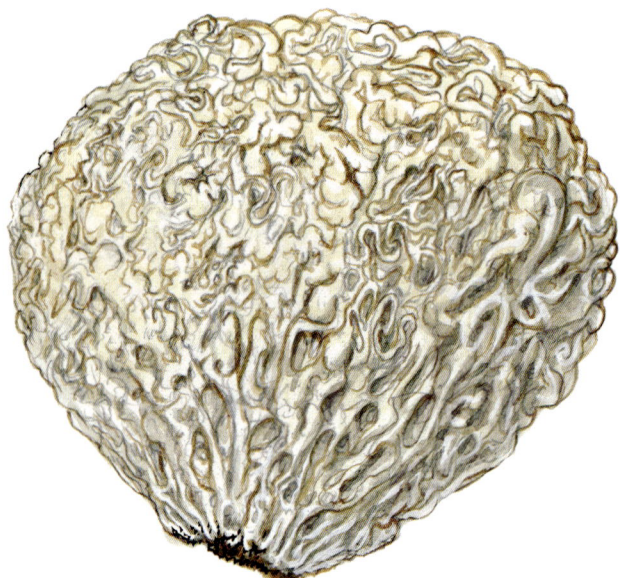

Bonfire
cauliflower

Peziza proteana f. sparassoides (Boud.) Korf

- ◉ **Mushroom in ball shape,** *irregular, lobed and contorted, whitish, able to reach 5 inches (12 cm) wide and 10 inches (25 cm) tall, looks like a Sparassis species.*
- ◉ **Flesh:** *fragile, membranous; sweet flavor and faint odor.*
- ◉ **Habitat:** *on old campfire sites that have been invaded by moss.*
- ◉ **When to observe:** *Autumn.*

Autumn newspapers in central Europe sometimes contain strange articles recounting tales of lucky mushroom hunters harvesting enormous "morels," complete with meticulously recorded measurements. Unfortunately, morels do not grow in autumn, and these purportedly fortunate hunters have usually found nothing more than this somewhat monstrous *Peziza* (also reported from the UK and from North America). It should also be noted that the bonfire cauliflower does not possess a stem (something all morels have) and that its membranous flesh is fragile, unlike the rubbery flesh of *Sparassis* species (see next page) it is sometimes confused with.

Did you know?

There are many mushrooms that prefer growing near charcoal ovens and other burned areas. This environmental particularity often shows up in their Latin name, which may include something like *carbonicolus,* "loves charcoal," *carbonarius,* "of charcoal," or *pyrophilus,* "loves fire." These mushrooms grow especially abundantly after forest fires.

Wood cauliflower

Sparassis crispa (Bull. : Pers.) Pers.

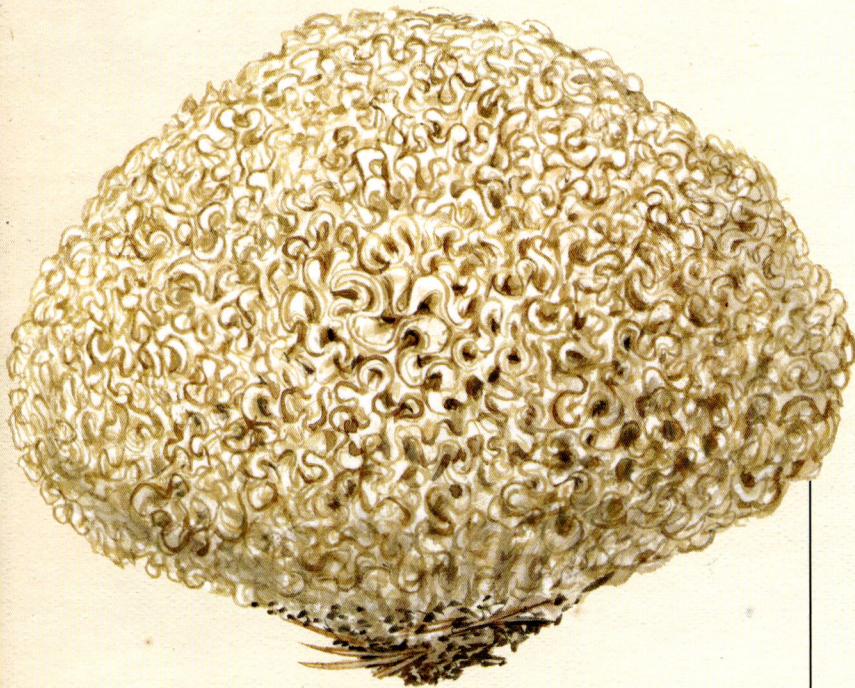

- ⊙ **Mushroom in "cauliflower" shape,** *4–16 inches (10–40 cm) in diameter, compact, made up entirely of short lobed folds resembling tagliatelle, cream then ochre or reddish-ochre.*
- ⊙ **Stem:** *absent.*
- ⊙ **Flesh:** *white, fairly substantial and a bit elastic; sweet flavor and pleasant odor.*
- ⊙ **Habitat:** *at the foot of dead or living conifers, especially under pine trees.*
- ⊙ **When to observe:** *Autumn.*

The wood cauliflower, reported from North America and from the UK, is not a very common mushroom but often grows for several years in the same place on old pine tree stumps or on the ground in relation with rotting roots. In addition to its habitat, it is known for its cauliflower-like appearance: up close, it seems to be made up of a multitude of membranous folds that have gotten a perm. Its close relative, *Sparassis brevipes* (not reported from the UK, but possibly reported from North America), prefers to grow on deciduous trees and does not have the same shape: it is much less frizzy, and its small membranous lamellae are shaped like fans and marked with concentric circles. Be careful not to confuse the wood cauliflower with the *"sparassis"* form of the bonfire cauliflower (see previous page). *S. spathulata,* most commonly found in pine forests and reported from the UK and North America, is very similar, as is *S. americana,* reported from hardwood forests in North America.

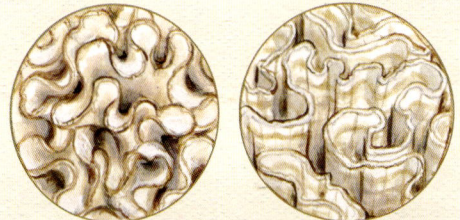

▲ Wood cauliflower ▲ S. brevipes

Did you know?

Sparassis are edible mushrooms, and the wood cauliflower is no exception. Once they have been washed and all of the impurities accumulated in their folds have been removed (not an easy task), they can be cooked and prepared like noodles.

Funiculus

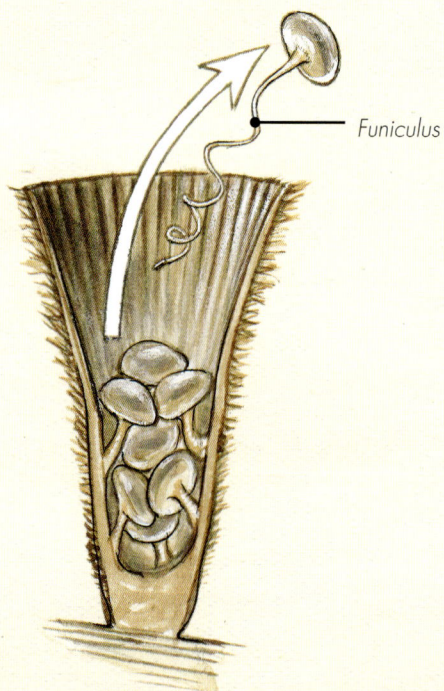

- ◉ **Mushroom in shape of "small nests"** from 0.5 to 1.5 cm tall and around 1 cm wide, external face spiky with brown bristles and a striped grayish internal face.
- ◉ Each **nest** is closed at first by a white or grayish film (the operculum) that tears to reveal small beige "eggs" that then turn gray (the peridioles).
- ◉ **Flesh:** very thin; faint odor and flavor.
- ◉ **Habitat:** on the ground on ligneous debris or on heavily decomposed pieces of wood.
- ◉ **When to observe:** Summer through autumn.

Fluted
bird's nest

Cyathus striatus (Huds. : Pers.) Pers.

▲ *Field bird's nest*

The fluted bird's nest, reported from North America and the UK, is very common but it takes a sharp eye to see it on the ground, because of how well it blends in with its environment.

When examined up close, though, its resemblance to a bird's nest filled with small eggs makes it easy to recognize. The spores are located inside of these eggs, which are projected out of the cup by a kind of spring called a funiculus when they are mature.

Did you know?

If you pay attention on your nature walks, you will find other mushrooms in this same group that also form "nests." The field bird's nest (*C. olla*), reported from the same geographic regions, forms cups that are covered in brownish felt on their external face and are gray and smooth on the internal face. *Crucibulum laeve*, reported from the same geographic regions, resembles the field bird's nest but is smaller, has a smooth external face, contains many small white peridioles, and grows in colonies on soil rich in organic debris.

▲ *Crucibulum laeve*

Candlesnuff
fungus

Xylaria hypoxylon (L. : Fr.) Grev.

- ⊙ **Mushroom 2–3 inches (5–8 cm) tall**, *appearing in two forms often blended together: arbuscules that are branched, black, and velvety at the bottom, covered in fine white dust at the top, and completely black clubs whose surface is marked with tiny black warts.*
- ⊙ **Stem:** *not very distinct, black.*
- ⊙ **Flesh:** *white, very leathery; flavor and odor not distinctive.*
- ⊙ **Habitat:** *on dead wood.*
- ⊙ **When to observe:** *All year long.*

If you pay attention on your walks in the woods, you cannot miss the candlesnuff fungus (reported from North America and the UK), no matter what time of year it is. It is certainly one of the most common mushrooms around, but its small black branches are not always easy to spot. Its flesh is extremely tough, and it is almost impossible to detach the mushroom from its substrate without breaking it. Other *Xylaria* species may be confused with the candlesnuff fungus, but they are much rarer: *X. filiformis* (reported from North America and the UK), for example, is very thin and grows on decomposing leaves.

Did you know?

The majority of mushrooms are able to reproduce in two different ways: the first is sexual reproduction, which involves the formation of sex cells (spores), and the second is called asexual reproduction and functions on the principle of cloning. The candlesnuff fungus illustrates both of these reproductive methods: the black and white arbuscules represent the asexual reproductive stage, and the black clubs are responsible for the sexual reproductive phase.

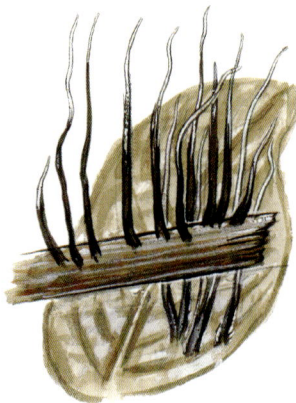

▲ *Xylaria filiformis*

coral or shrub shape | 286

Dead
man's fingers

Xylaria polymorpha (Pers.) Grev.

- ⊙ **Mushroom in club shape,** black, 2–4 inches (5–10 cm) tall, very leathery, matte, smooth surface with wrinkles or tiny black warts; often stains the fingers black.
- ⊙ **Stem:** inside the extension of the club, black.
- ⊙ **Flesh:** white, very tough; flavor and odor not distinctive.
- ⊙ **Habitat:** on dead wood.
- ⊙ **When to observe:** All year long.

The dead man's fingers, reported from North America and the UK, is probably a little less common than the candlesnuff fungus (previous page) but is just as easy to recognize. Its extremely leathery black clubs solidly grafted onto dead wood are typical. If you cut one in half lengthwise with a sharp knife, you will see the white flesh and small spherical cavities around the edge of the club that open to the exterior through a microscopic orifice. In these cavities, which mycologists call "perithecia," black spores are formed and liberated at maturity through the apical opening known as an ostiole.

Did you know?

It bears repeating that the shape of a mushroom is often a poor criterion for proving relationships between various groups. Even though the dead man's fingers looks like a small club-shaped *Clavaria*, it belongs to the large ascomycete group that contains morels, *Peziza*, and truffles.

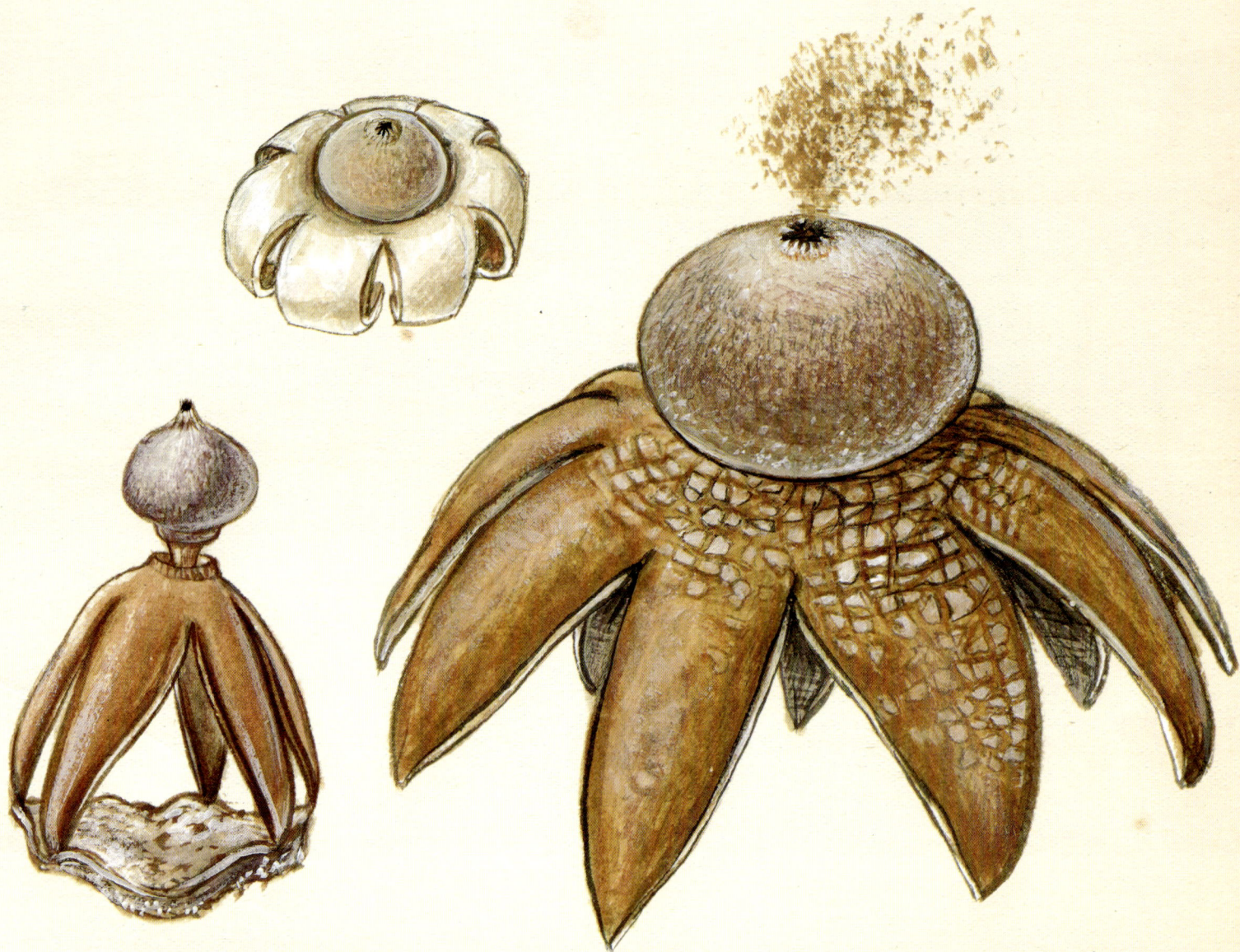

- **Mushroom in hard ball shape at first** reaching just over 1 inch (3 cm) wide, then opening into a star formation; the star's branches close during dry weather; central ball is fairly soft, opening at the top via an irregular hole.

- **Flesh:** white then brown; sweet flavor and faint rubbery odor.
- **Habitat:** under deciduous and coniferous trees in sandy soil.
- **When to observe:** Springtime through late autumn.

Barometer
earthstar

Astraeus hygrometricus (Pers. : Pers.) Morgan

The gasteromycetes

At one time, all mushrooms that formed their spores in a stomach-like sac that opened at maturity were classified as gasteromycetes, from the Greek gastêr, *gastros, "stomach," and* mukês, *"mushroom." Mycologists noticed, however, that this morphology had appeared over the course of evolution in several mushroom groups that are not directly related. For this reason, the gasteromycete group is no longer recognized and the mushrooms that once were part of it have been divided based on their various biological origins. The Astraeus, for instance, are now grouped closer to the* Scleroderma *(p. 277) in the large bolete group.*

The barometer earthstar, reported from North America and the UK, is a small mushroom that you may occasionally see on the forest floor in sandy soil. Its colors are difficult to distinguish from the leaves it grows on, especially in dry weather when its branches are closed over the central "ball" containing the spores. Quite a few *Geastrum* species resemble it because of their star shape, but the orifice in their central ball, the ostiole, is typically well-defined and more regular.

Warning

While the barometer earthstar is considered an uninteresting mushroom for culinary purposes, it has apparently been responsible for rather severe poisonings in animals, dogs in particular. It is therefore logical to believe that this mushroom is toxic. At the same time, however, it produces molecules that are capable of acting against certain kinds of cancer cells and stimulating the immune system.

Did you know?

To the best of our scientific knowledge, only five *Astraeus* species exist in the world. The barometer earthstar is the only one present in Europe. *Astraeus pteridis* (reported from North America) and another species that does not yet have a name are found in the United States, and *A. odoratus* and *A. asiaticus* have been observed in Asia. All *Astraeus* are mycorrhizal and associate with the trees around them.

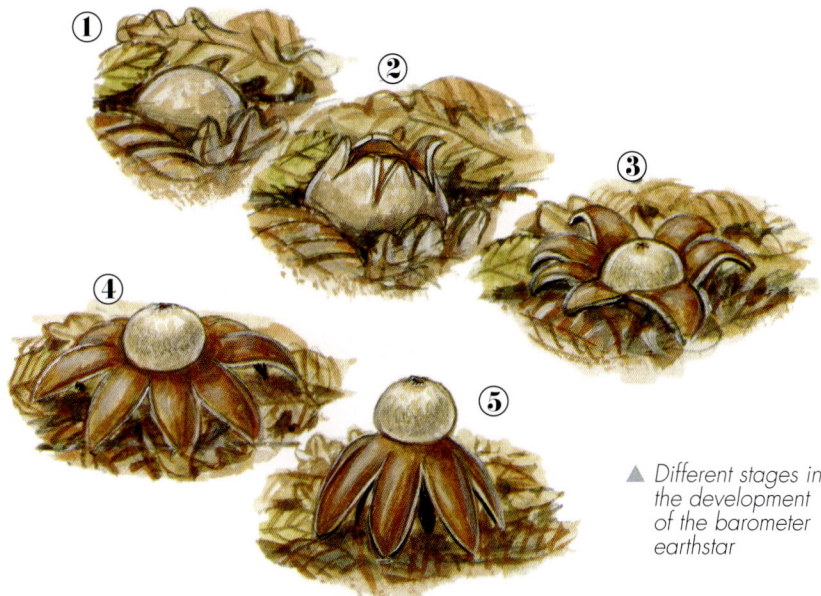

▲ *Different stages in the development of the barometer earthstar*

- ◉ **Mushroom in club shape with blunt, rounded top,** can reach 8–10 inches (20–25 cm) in height, yellow then ochre-yellow, reddens as it grows or if rubbed; spore print cream.
- ◉ **Flesh:** white, fairly thick; reddens when cut; flavor both sugary and bitter, faint odor.
- ◉ **Habitat:** especially under beech and oak trees.
- ◉ **When to observe:** Summer into autumn.

Giant club

Clavariadelphus pistillaris (L. : Fr.) Donk

This almost impossible to confuse mushroom is not very common but is reported from North America and the UK. We sometimes see it in small groups in older and relatively undisturbed deciduous forests. *C. truncatus* (reported from North America but not the UK) has a similar appearance, but the top of its club—as the name indicates—is truncated and almost flat, as if the giant club had been inadvertently chopped by a lawnmower blade.

Did you know?

Two other *Clavariadelphus* in addition to *C. truncatus* exist in Europe, and both are distinctly smaller than the giant club. *C. helveticus* (not reported from the UK or North America), creamy white to pale ochre, grows under fir and spruce trees, while *C. ligula* (reported from North America and the UK) does not redden when cut, or only faintly, and grows under a variety of trees. *C. americanus*, reported from North America, is smaller than *C. pistillaris*, and is more frequently found in conifer forests. In colder, more northern conifer forests of North America and the UK, is *C. sachalinensis*, which is similar to *C. ligula*.

▲ *Clavariadelphus pistillaris*

▲ *Clavariadelphus truncatus*

▲ *Clavariadelphus helveticus*

▲ *Clavariadelphus ligula*

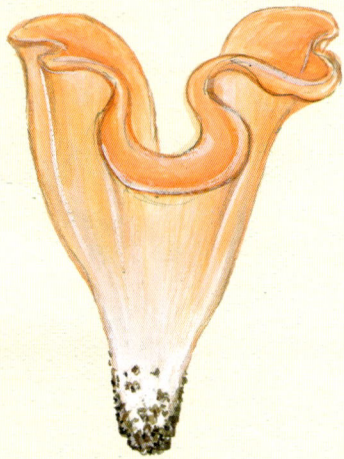

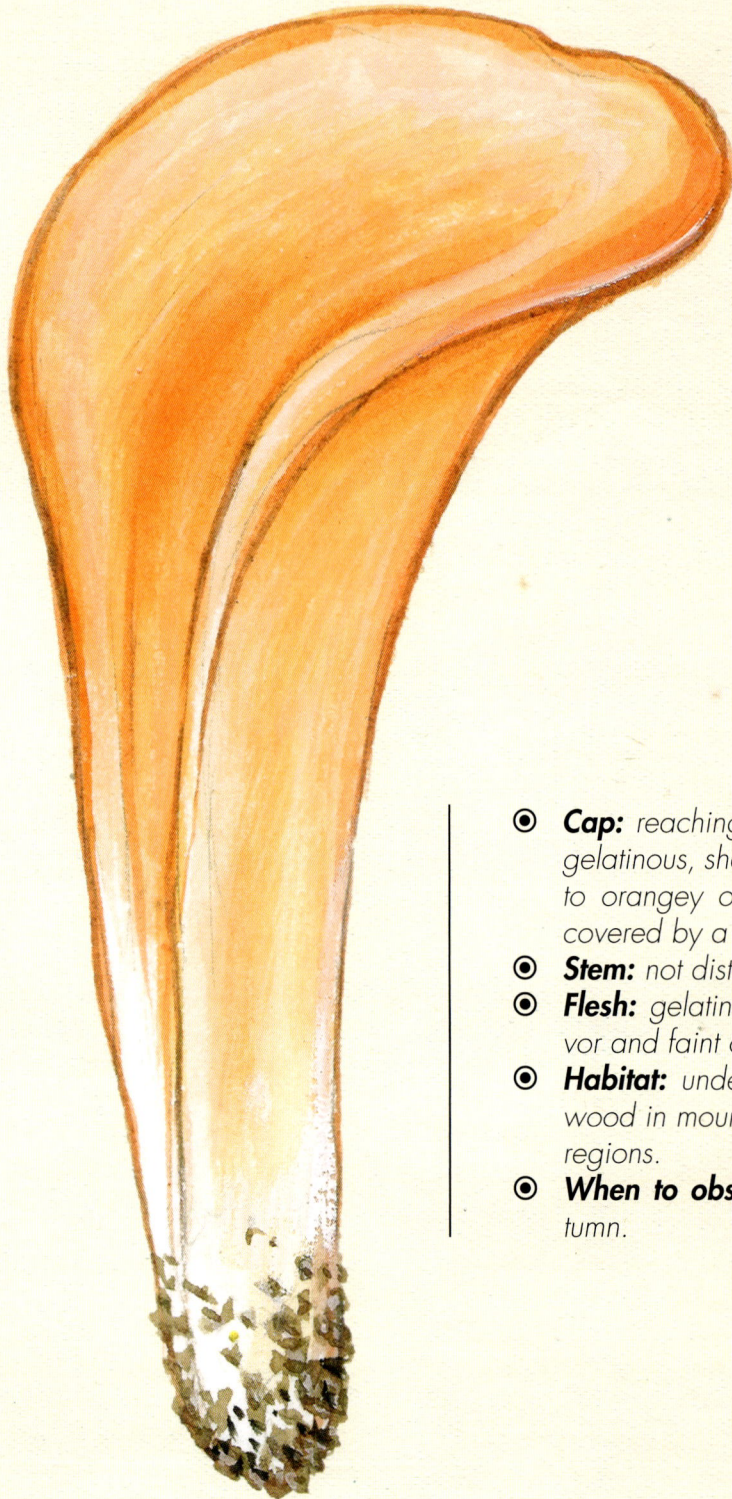

- ◉ **Cap:** *reaching 6 inches (15 cm) tall, gelatinous, shaped like a tongue, pink to orangey or reddish-pink, partially covered by a whitish bloom.*
- ◉ **Stem:** *not distinct from the cap.*
- ◉ **Flesh:** *gelatinous and soft; sweet flavor and faint odor.*
- ◉ **Habitat:** *under conifers or on rotting wood in mountainous areas or colder regions.*
- ◉ **When to observe:** *Summer and autumn.*

Salmon
salad

Tremiscus (*Guepinia*) *helvelloides* (DC. : Fr.) Donk

The salmon salad is reported from North America and the UK and is impossible to mistake for any other mushroom. Its habitat in and around mountain conifer forests (or in colder regions), its color, and its gelatinous texture are characteristic features. Only the jelly tooth (*Pseudohydnum gelatinosum*) has a similar consistency, but it is whitish or grayish and the underside of its cap is covered with fine translucent teeth. Certain *Tremella* species also have this gluey appearance, but they are yellow or brown and have a much different shape. While *Otidea onotica*, reported from the same geographic regions, has the same general shape and color, though more yellow, it has brittle, rather than gelatinous, flesh.

▲ *Jelly tooth*

Warning

While the salmon salad is usually considered edible, it lacks flavor and its soft flesh will not be to everyone's liking. Some mushroom lovers suggest eating it raw in salads or pickled in vinegar and herbs. Note that it becomes leathery and indigestible as it ages.

Did you know?

Like many mushrooms, the salmon salad produces large molecules called polysaccharides. Researchers have discovered that these substances have an inhibitive effect on the growth of certain cancers in mice.

A difficult baptism

It was Dutch botanist Nikolaus von Jacquin who granted the salmon salad its first Latin name, Tremella rufa, in 1778. Unfortunately for von Jacquin and his descendants, he was a contemporary of renowned naturalist Carl Linnaeus, whose work and the names within it became the references used by the International Code of Botanical Nomenclature to govern the use of Latin names. Instead of using the name von Jacquin had created, Linnaeus kept the name Tremella helvelloides (since changed to Tremiscus helvelloides) proposed by Jean-Baptiste de Lamarck in his great text entitled Flore française ou description succincte de toutes les plantes qui croissent naturellement en France ("French Flora or Succinct Description of all Plants Growing Naturally in France"). As a result, the name Tremella rufa is now forgotten.

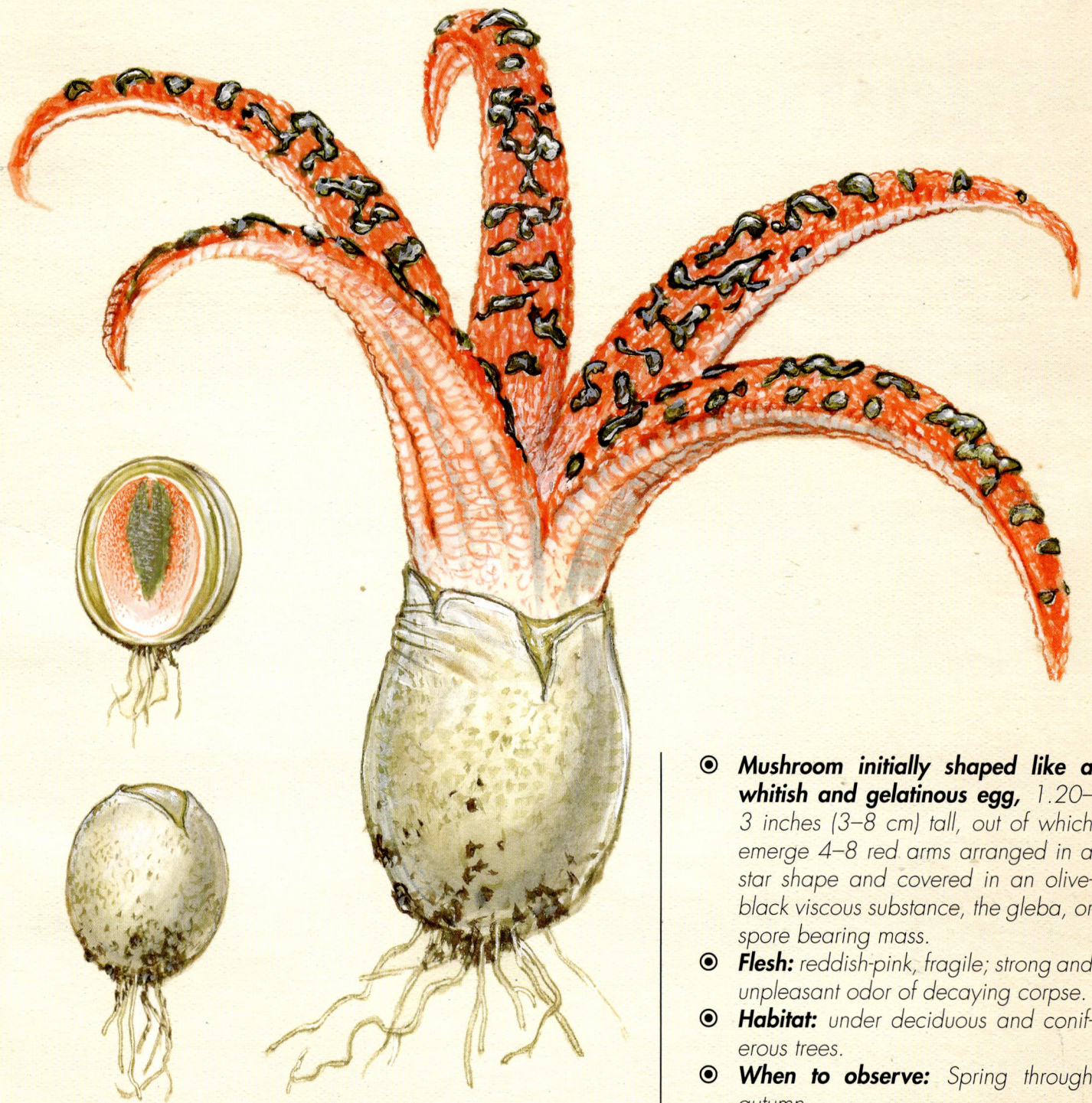

- ⊙ **Mushroom initially shaped like a whitish and gelatinous egg,** 1.20–3 inches (3–8 cm) tall, out of which emerge 4–8 red arms arranged in a star shape and covered in an olive-black viscous substance, the gleba, or spore bearing mass.
- ⊙ **Flesh:** reddish-pink, fragile; strong and unpleasant odor of decaying corpse.
- ⊙ **Habitat:** under deciduous and coniferous trees.
- ⊙ **When to observe:** Spring through autumn.

Devil's
fingers

Clathrus archeri (Berk.) Dring

The devil's fingers, reported from the UK and from North America, never misses a chance to attract the attention of passerby, making its presence known not only with its shape and color—which are already unique—but also its repugnant odor, which announces itself from several yards away before you have even spotted the mushroom. The blackish and viscous substance covering its branches is especially odorous because it contains the spores; flies and other insects that are attracted by what they find an appetizing scent and leave with their legs covered in spores, in this way ensuring the dispersal of the species. *Pseudocolus fusiformis*, reported from eastern North America, is similar, but the tips of its fingers are fused.

Warning

The devil's fingers is probably not toxic, but whoever decides to con-

Invaders

The devil's fingers, like the red cage (p. 296), did not originate in temperate climates. Both mushrooms come from the other side of the world, more specifically the Australasian region. They seem to have found their way to us with the help of the international wool trade: their spores, trapped in the coats of sheep grown in the Southern Hemisphere, were liberated in Europe by the water used to wash the wool. The spores found European countries perfectly suitable, and these mushrooms expanded in number. They are sometimes considered an invasive species, but their negative impact on native species is difficult to prove.

sume it will have to be brave (or starving). On the other hand, if you are considering harvesting its eggs when they are still closed for the pleasure of watching them develop in your yard, avoid transporting them long distances or leaving them for hours in your car. The arms can emerge and develop very quickly, and despite your best efforts you may end up "perfuming" your car and forever remembering this regrettable oversight . . .

Did you know?

In addition to the *Clathrus* mentioned here, there are other mushrooms that have settled after arriving from elsewhere. Some happily while away their days without being noticed, but others have caused ecological catastrophe. In the early 1970s, a small mushroom named *Ophiostoma*

ulmi arrived in Europe; within only a few years it had destroyed almost every elm tree (the disease it causes is called graphiosis, or Dutch elm disease). These once majestic trees no longer exist as mature trees, only as scrawny specimens that rarely reach adulthood. The small, deadly ascomycete arrived in 1930 in central North America and quickly spread. Today, elms survive in some wet habitats, but never reach majestic old age.

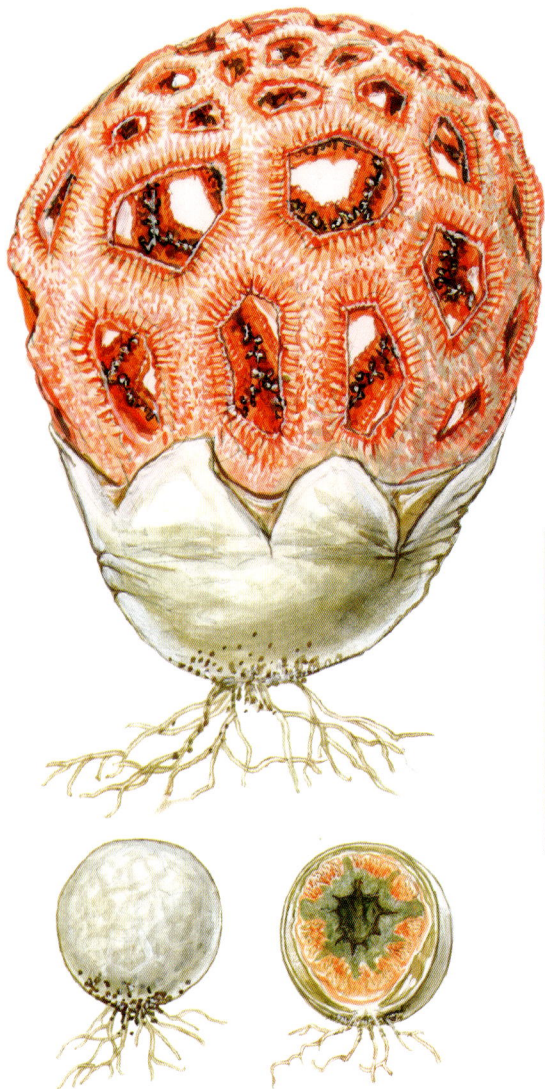

Red
cage

Clathrus ruber Micheli : Pers.

- ◉ **Mushroom initially shaped like a whitish gelatinous egg**, 1.20–2 inches (3–5 cm) tall, out of which emerges a sort of red mesh cage that can reach 6 inches (15 cm) in diameter with an interior face covered by an olive black viscous substance.
- ◉ **Flesh:** reddish-pink, fragile; strong and unpleasant odor of decaying corpse.
- ◉ **Habitat:** under deciduous and coniferous trees.
- ◉ **When to observe:** Spring through autumn.

The red cage, reported from the UK and North America, is much less common than its close relative the devil's fingers (p. 295). It typically grows in warm regions, but is not completely absent in northern regions. It is easily recognized by its cage shape, even though its cage is so fragile that it often displays irregularities and pieces that have broken during growth. It may be confused with *Colus hirundinosus* (reported from North America but not from the UK), a species that shares characteristics with both the devil's fingers and the red cage. *C. hirundinosus* forms arms like tentacles that are red like the devil's fingers, but these arms are connected at the top by a small mesh that resembles the red cage.

▲ *Red cage*

▲ *Colus hirundinosis*

Did you know?

While *Clathrus* mushrooms are relatively rare in Europe and North America, they are far more abundant in their preferred tropical and equatorial regions. In these parts of the world, they assume a variety of shapes that sometimes resemble *Clathrus* in parts of Europe but are more often astonishingly different, resembling octopuses or brightly colored coral.

Dog
stinkhorn

Mutinus caninus (Huds. : Pers.) Fr.

- ◉ **Mushroom initially in the shape of a white gelatinous egg,** *fairly elongated and extended by a few threads of mycelium, out of which emerges a fragile cylindrical "stem" that can reach 4 inches (10 cm) tall, whitish or pale orange, terminating in an orangey or reddish cone that is covered by a blackish-olive viscous substance.*
- ◉ **Flesh:** *fragile, pitted with many small holes; sweet flavor and fairly strong and unpleasant fecal odor.*
- ◉ **Habitat:** *especially in deciduous forests, often near dead wood.*
- ◉ **When to observe:** *Summer and autumn.*

The dog stinkhorn, reported from North America and the UK, is hard to miss because of its rather suggestive shape, and its repugnant odor. It is the only *Mutinus* species native to Europe, but in some regions you may encounter two other species originally from North America that have become naturalized: *M. elegans*, not reported from the UK, has a salmon to orangey-red "stem" the same color as the head, with an olive gleba, and *M. ravenelii*, reported from the UK, has a pointier head than the dog stinkhorn that does not contrast as sharply with the partially colored stem. The *Mutinus* are close relatives of the stinkhorn (p. 299).

Did you know?

Despite its less than encouraging appearance and foul odor, the dog stinkhorn is considered edible in some places, particularly in the eastern United States. Lovers of this mushroom advise not waiting for it to develop and instead to only consume the eggs, peeled and fried. If you have nothing else more appetizing to eat, that is.

- **Mushroom initially in the shape of a white gelatinous egg,** *1.20–3 inches (3–8 cm) tall and elongated by a few threads of mycelium, out of which emerges a fragile cylindrical "stem" that can grow 6 inches (15 cm) tall, whitish, terminating in an alveolate conical head positioned like a thimble, covered at first by a blackish-olive viscous substance.*
- **Flesh:** *fragile, pitted with many small holes; sweet flavor and very strong and unpleasant odor of decaying corpse.*
- **Habitat:** *in coniferous and deciduous forests.*
- **When to observe:** *Summer and autumn.*

Stinkhorn

Phallus impudicus L. : Pers.

Like its close relatives the *Clathrus* (pp. 295–296), the stinkhorn can be detected from far away by the strong odor emanating from the blackish-olive viscous substance covering its "cap." It is this odor that, by attracting flies and other necrophagous insects, allows the tiny spores to be transported great distances, carried on the legs or in the digestive system of these small creatures. *P. impudicus* is reported from North America, and the UK. *P. duplicatus*, reported most commonly from eastern North America, is very much like the stinkhorn in that it has green gleba and a pitted head, but a lace-like veil hangs below its head.

hadriani (also reported from the UK and North America). It can be differentiated from the stinkhorn by the vertical pleats on its egg and the way its egg turns pink or purplish-pink when exposed to air or bruised. It grows in warm regions of southern Europe and on sandy ground, especially on dunes. It should also be noted that, on rare occasions, the stinkhorn develops a kind of lacy veil along the inferior border of its "head" called an indusium: when this happens, the stinkhorn is in its *togatus* form.

Did you know?

If you find a gelatinous egg that clearly belongs to the stinkhorn family (the *Phallaceae*), how do you know if it comes from genus *Phallus* or genus *Clathrus* (pp. 295–296)? Simply cut

it in half: if the cut reveals a stem in formation, it is a *Phallus* egg. If, on the other hand, you notice portions of red flesh, you have found a *Clathrus* egg. Note that *Phallus* species are edible in the egg stage; the stem has a brittle and gristly texture but a pleasant hazelnut flavor.

▲ *Phallus hadriani*

Warning

There is only one other *Phallus* species in mainland Europe, *Phallus*

The cousin from the tropics

The stinkhorn sometimes appears with a small veil just below the base of the mushroom's "head." Among Phallus indusiatus specimens (reported from southern North America) in tropical and equatorial regions, however, this little veil becomes a veritable skirt that is highly developed, elegant, and sometimes colorful. The bridal veil stinkhorn, as it is sometimes called in English, has been cultivated in China since the 1980s and is sold dried for use in a variety of traditional dishes. It is said that Empress Dowager Cixi of the Qing dynasty served it in a soup as part of her sixtieth birthday celebration.

Recipe Book

Mushroom sauce

Preparation: 10 min   Cooking time: 15 min

Serves 6

- ⊙ 5 ¼ ounces (150 g) mixed wild mushrooms
- ⊙ 2 shallots
- ⊙ ½ bunch chervil
- ⊙ 1 ¾ tablespoons (25 g) butter
- ⊙ 6 ¾ tablespoons (10 cl) dry white wine
- ⊙ ⅔ cup (15 cl) chicken stock
- ⊙ 1 cup (25 cl) heavy cream
- ⊙ Salt and freshly ground pepper

1 Clean the mushrooms. Peel and chop the shallots. Remove the chervil leaves from the stems.

2 In a pan, melt the butter then add the shallots and mushrooms, cooking until the shallots are translucent.

3 Pour in the white wine and chicken stock. Reduce the liquid by one quarter. Add the cream and reduce another half. Salt and pepper to taste, then add the chervil. Serve immediately.

Cep soup with
parmesan crisps

Preparation: 20 min (not including broth)   Cooking time: 25 min

Serves 6

- ⊙ 1 cup (100 g) freshly grated parmesan
- ⊙ 18 ounces (500 g) ceps
- ⊙ 2 shallots
- ⊙ 2 tablespoons (30 g) butter
- ⊙ 32 ounces (1 liter) chicken stock or 1 organic chicken bouillon cube + 32 ounces (1 liter) of water
- ⊙ ¾ cup (20 cl) heavy cream
- ⊙ Salt

Equipment
- ⊙ Blender
- ⊙ Baking sheet
- ⊙ Parchment paper
- ⊙ Skimmer

1 Preheat the oven to 425°F (220°C, thermostat 7). On the baking sheet lined with parchment paper, spread the parmesan into disc shapes around 4 inches (10 cm) wide. Bake for around 5 minutes until the parmesan melts. Leave to cool so the crisps can harden.

2 Scrub the ceps and mince them. Peel and mince the shallots. In a pot, cook the shallots in the melted butter with a pinch of salt. Add the ceps and continue cooking for 5 minutes, stirring regularly. Pour in the stock. Cover and bring to a boil, then leave to cook on low heat for 15 minutes. Drain the ceps quickly using a skimmer and set the broth aside.

3 Stir the ceps into the cream and pour in the broth little by little until you reach a velvety consistency. Serve hot with parmesan crisps.

Parmesan
cep carpaccio

Preparation: 15 min

Serves 4

- ◉ 10 ½ ounces (300 g) small firm ceps
- ◉ 3 ½ ounces (100 g) fresh parmesan
- ◉ 4 ¼ tablespoons olive oil
- ◉ 1 ⅔ tablespoons reduced balsamic vinegar
- ◉ Fleur de sel and freshly ground pepper

Equipment
- ◉ Mandoline
- ◉ Vegetable peeler

1 Scrub the ceps and cut off the earthy part of the stems. Slice them thinly using a mandoline or very sharp knife. Create parmesan shavings using a vegetable peeler.

2 Divide the ceps onto four plates, sprinkle over the parmesan shavings, then drizzle with olive oil and reduced balsamic vinegar. Season with fleur de sel and pepper and serve immediately.

Sweetbreads à la crème
with mushrooms

Preparation: 30 min   Cooking time: 35 min

Serves 4

- ◉ 1 ¼ pounds (600 g) sweetbreads, ready to cook
- ◉ 2 shallots
- ◉ 14 ounces (400 g) ceps
- ◉ 2 teaspoons butter
- ◉ 1 pinch of thyme, fresh if possible
- ◉ 3 ⅓ tablespoons (5 cl) cognac
- ◉ ¾ cup (20 cl) stock
- ◉ ¾ cup (20 cl) crème fraîche
- ◉ 1 bunch chives
- ◉ Peanut oil
- ◉ Salt and freshly ground pepper

1 Place the sweetbreads in a saucepan, cover with cold water, and bring to a boil. Let cook 1 minute and drain immediately.

2 Peel and chop the shallots. Cut the ceps if they are large, otherwise leave them whole.

3 In a pot, cook the sweetbreads seasoned with salt and pepper over low heat in the butter and a little oil. Wait around 10 minutes. Add the shallots and mushrooms and allow to cook 5 minutes. Stir occasionally. Add the thyme.

4 Pour in the cognac and bring to a boil, then add the stock and the crème fraîche. Season with salt and pepper. Cook for around 20 minutes covered over low heat. Turn over the sweetbreads now and then. Add the chopped chives when the pot has been removed from the heat and serve piping hot.

Suggestion
Avoid overcooking sweetbreads because they can become slightly rubbery.

Option
You can replace the ceps with other mushrooms.

Forester's
pizza

Preparation: 30 min   Cooking time: 15 min

Serves 4

- ◉   1 pizza dough
- ◉   1 ball fresh mozzarella
- ◉   1 garlic clove
- ◉   10 sprigs of parsley
- ◉   ¾ tablespoon (10 g) butter
- ◉   9 ounces (250 g) button mushrooms
- ◉   9 ounces (250 g) forester's blend
       (*Pleurotus*, yellow boletes, chanterelles)
- ◉   2 tablespoons thick crème fraîche
- ◉   1 cup (100 g) grated emmental
- ◉   Salt and freshly ground pepper

Equipment
- ◉   Baking tray used for pastries or pizza

1 Preheat the oven to 390°F (200°C, thermostat 6–7). Roll out the pizza dough and place it on the baking tray.

2 Drain the mozzarella and cut it into cubes. Peel and mince the garlic. Chop the parsley.

3 In a pan, melt the butter and cook the button mushrooms and forester's blend with the garlic and parsley. All of the water should be absorbed. Season with salt and pepper and set aside.

4 Spread the crème fraîche over the dough and arrange the mushrooms. Arrange the mozzarella cubes and sprinkle with grated emmental. Bake for 15 minutes.

Tagliatelles and chanterelles in
chive cream sauce

Preparation: 25 min   Cooking time: 20 min

Serves 4

Pasta
- 14 ounces (400 g) mushroom-flavored tagliatelle

Sauce
- 18 ounces (500 g) chanterelles
- 1 onion
- 1 garlic clove
- 1 bunch chives
- 3 ½ ounces (100 g) hazelnuts
- 1 drizzle of olive oil
- 1 ½ tablespoons butter
- ¾ cup (20 cl) thick crème fraîche
- 1 cup (100 g) freshly grated parmesan
- Salt and freshly ground pepper

Cooking
- 48 ounces (1.5 liters) chicken broth

1 Prepare the sauce. Scrape the stems of the chanterelles with the tip of a knife, then wash the mushrooms in several baths of clear water and drain.

2 Peel and mince the onion. Peel the garlic, remove the sprout, and crush the clove in a garlic press. Chop the chives. Grill the hazelnuts in the oven for 5 minutes. Allow them to cool and break them into large chunks using the flat side of a knife.

3 In a pan, heat a drizzle of olive oil and cook the chanterelles for 3 minutes. Remove them and discard the cooking liquid. In the same pan, melt the butter and fry the chanterelles for 3–4 minutes. Add the onion and garlic. Season with salt and pepper and cook 1 minute. Add the chopped chives, the crushed hazelnuts, and the crème fraîche. Keep warm.

4 In a pot, boil the chicken broth. Plunge the tagliatelle into the bubbling broth. As soon as they are cooked *al dente*, drain and pour them into the pan with the rest of the ingredients. Stir together for 30 seconds. Serve immediately with grated parmesan.

Foie gras and
morel terrine

Preparation: 30 min   Cooking time: 30 min   Refrigeration: 1 hour + 2–3 days

Serves 2

- 1 raw duck foie gras weighing about 14 ounces (400 g)
- 2 tablespoons (3 cl) Arbois wine
- 1 teaspoon (5 g) salt
- A dozen rounds of freshly ground pepper
- 1 ¾ ounces (50 g) morels
- 1 ½ tablespoons (20 g) butter

Equipment
- 2 mini terrine molds

1 Prepare and devein the foie gras. Season it with the salt, pepper, and Arbois wine and refrigerate for at least 1 hour.

2 During this time, clean the morels by removing the earthy stems. Cut the mushrooms in half depending on their size. Sauté the morels for 5 minutes in the butter then set aside.

3 Place one quarter of the foie gras in each of the mini terrine molds. Cover with the morels, then add the rest of the foie gras, packing the pieces close together to eliminate as much air as possible. Put the covers on the molds and cook for 20 minutes using a bain-marie in an oven preheated to 300°F (150°C, thermostat 5).

4 After removing the molds from the oven, take off the covers and weigh down a piece of cardboard the same size as the terrine and wrapped in aluminum foil on top of each serving. Keep for 2–3 days in the refrigerator before tasting.

Chicken and
mushroom samosas

Preparation: 30 min   Cooking time: 8–10 min

For 20 samosas

- ◉ 7 ounces (200 g) chicken breast
- ◉ 14 ounces (400 g) cleaned hedgehog mushrooms
- ◉ A few sprigs of flat-leaf parsley
- ◉ 1 shallot
- ◉ 2 tablespoons olive oil
- ◉ 2 tablespoons crème fraîche
- ◉ 10 sheets of brik pastry or filo dough
- ◉ 32 ounces (1 liter) frying oil
- ◉ Salt and freshly ground pepper

1 Cut the chicken into small pieces. Slice the mushrooms into slivers and chop the parsley. Peel and finely mince the shallot.

2 In a pan, sauté the shallot in the olive oil. Add the mushrooms and parsley. Stir. Add the pieces of chicken and continue cooking until the chicken is cooked through. Add the crème fraîche. Mix well and season to taste.

3 Prepare the brik pastry sheets: cut each sheet in half, removing the rounded section on the bottom to obtain straight-edged strips; if you are using sheets of filo dough, cut them into strips.

4 Spoon 1 tablespoon of filling onto one end of the strip and fold the dough over carefully to form a triangle. Don't hesitate to press down as you do this to keep the samosas from opening during cooking. Fold the last section over again to complete the triangle by tucking the last section inside the samosa.

5 Heat the frying oil in a deep pan. Quickly plunge the samosas into the boiling oil until they are golden brown.

Vegetable
tempura

Preparation: 10 min   Cooking time: 10 min

Serves 6

- ⊙ 12 baby leeks
- ⊙ 12 baby carrots
- ⊙ 12 mushrooms (*Pleurotus*, button mush-rooms, etc.)
- ⊙ 1 cup (130 g) flour
- ⊙ 3 ¾ tablespoons (30 g) cornstarch
- ⊙ 1 egg
- ⊙ ¾ cup (20 cl) chilled sparkling water
- ⊙ 4 ice cubes
- ⊙ Frying oil
- ⊙ Salt and freshly ground pepper

Equipment
- ⊙ Fryer
- ⊙ Vegetable peeler

1 Clean the baby leeks: cut off the base, remove the outer leaves, and slice off some of the green portion. Peel the carrots. Clean all of the vegetables and slice the mushrooms.

2 In a bowl, mix together the flour, cornstarch, salt, and pepper. Add the egg then the sparkling water. Stir briefly and add the ice cubes to keep the tempura batter cool.

3 Plunge the vegetables into the batter then into the hot oil at 350°F (180°C) for a few minutes. Drain on a paper towel and eat with bamboo chopsticks.

Mini chanterelle pâtés
with red onion and basil

Preparation: 1 hour   Cooking time: 40 min

Serves 10

- ◉ 5 ½ pounds (2.5 kg) chanterelles
- ◉ 1 red onion
- ◉ 3 tablespoons olive oil
- ◉ 3 eggs
- ◉ 1 tablespoon cornstarch
- ◉ 2 tablespoons crème fraîche
- ◉ 2 teaspoons (10 g) salt
- ◉ 4 pinches Szechuan pepper
- ◉ 4 pinches ground cardamom
- ◉ 2 pinches freshly ground black pepper
- ◉ 1 large bunch basil
- ◉ Butter for the molds

Equipment
- ◉ Mixer
- ◉ 10 small individual terrine molds

1 Remove the mushroom stems to eliminate any traces of dirt or sand if necessary. Wash if needed and dry. Peel and chop the red onion.

2 In a pan, sauté the onion in olive oil over medium heat for 2 or 3 minutes. Add the chanterelles and cook for about 15 minutes, covered, stirring often. Drain them over a bowl and press to squeeze out any liquid. Then allow them to cool. Keep the liquid that drained off and boil it in a pot to obtain 1.5–2 tablespoons of highly concentrated liquid. Set aside.

3 Finely chop three quarters of the cooked mushrooms, then place them in a bowl with the rest of the whole mushrooms. Add the eggs, cornstarch, crème fraîche, salt, pepper, spices, the mushroom cooking liquid, and the chopped basil. Mix well.

4 Preheat the oven to 285°F (140°C, thermostat 4–5). Butter the molds and fill with the mixture. Place them in the oven to cook for around 40 minutes, depending on the size of the molds. Serve the pâtés hot in their molds and eat with a spoon.

Sautéed ceps
with shallots

Preparation: 25 min   Cooking time: 30 min

Serves 4

- ◉ 2 ⅔ pounds (1.2 kg) ceps
- ◉ 5 shallots
- ◉ 1 ½ tablespoons (20 g) butter, divided
- ◉ 3 ½ tablespoons (5 cl) vermouth (Noilly Prat)
- ◉ 1 cup (25 cl) crème fraîche
- ◉ 2 pinches grated nutmeg
- ◉ A few sprigs of parsley
- ◉ A few drops of lemon juice
- ◉ Salt and freshly ground pepper

1 Wipe off the ceps and remove any pieces of twigs or grass. If they are frozen, let them thaw. Peel and finely chop the shallots.

2 Cut the ceps in thick slices, then sauté them in a pan with half of the butter for 10 minutes.

3 In a pot, cook the shallots in the rest of the butter for 2 or 3 minutes. Add the ceps and stir. Pour in the vermouth and allow to boil for 1 minute before adding the crème fraîche, nutmeg, and a little salt and pepper. Allow to simmer for around 20 minutes.

4 Remove from heat, add salt and pepper if needed, then finish the dish by adding the chopped parsley and a few drops of lemon juice.

⫙ Option
This recipe also works with chanterelles or button mushrooms.

Asparagus
mushroom casserole

Preparation: 25 min   Cooking time: 20 min

Serves 4

- ⊙ 2 ¼ pounds (1 kg) asparagus
- ⊙ 1 carrot
- ⊙ 1 onion
- ⊙ 3 ½ ounces (100 g) scarletina bolete
- ⊙ 3 ½ ounces (100 g) smoked bacon cubes
- ⊙ 1 ½ tablespoons (20 g) butter, divided
- ⊙ 1 cup (25 cl) chicken broth
- ⊙ 1 bunch chervil
- ⊙ Salt and freshly ground pepper

Equipment
- ⊙ Cooking twine
- ⊙ Skimmer
- ⊙ Vegetable peeler

1 Peel the asparagus with a peeling knife then rinse under running water.

2 Chop the asparagus stalks in half and tie them in two bunches with the cooking twine. Cook for 7 or 8 minutes in a pot of salted boiling water.

3 Peel the carrot and onion. Clean the mushrooms. Roughly chop the vegetables. Remove the asparagus from the water using a skimmer, remove the twine and set aside.

4 In a saucepan, cook the vegetables and smoked bacon cubes in half of the butter for 2 or 3 minutes. Then add the asparagus, cut into thirds. Add salt and pepper and allow to cook for 2 minutes.

5 Pour in the chicken broth, cover, and cook for 10 minutes. Remove from heat, add the rest of the butter and chopped chervil. Serve hot, with the juices.

Cep
platter

Preparation: 25 min   Cooking time: 20 min

For 4 platters

Onion confit
- ⊙ 2 bunches of green onions
- ⊙ 5 tablespoons honey
- ⊙ 3 tablespoons red wine vinegar
- ⊙ Olive oil
- ⊙ Salt and freshly ground pepper

Mushrooms
- ⊙ 5 fresh ceps, divided
- ⊙ 4 ¼ ounces (120 g) marinated ceps
- ⊙ 1 garlic clove, chopped
- ⊙ A few sprigs of parsley
- ⊙ 2 pats of butter
- ⊙ Olive oil

Wine
- ⊙ Côte-Rôtie

1 Wash the onions and chop them roughly, keeping half of their green stalks. In a small pot or pan, cook with the olive oil for around 10 minutes. Add the honey, salt, and pepper, and cook over steady heat until they begin to brown. Add the vinegar, cook for another 2 minutes, remove from heat, and cool. Set aside.

2 Cut four ceps in slices that are not too thin. Cook in a pan with hot butter and a little oil. Stir gently. When they begin to brown, add the garlic and a little chopped parsley, stir, cook for a few moments and set aside. Cut the remaining fresh cep in thin slices.

3 On the platters, evenly distribute the onion confit, the drained marinated ceps, the sautéed ceps, and the slices of raw cep. Sprinkle salt and pepper over the raw mushroom slices and drizzle with a little olive oil. Serve with wine.

Option
Tweak this recipe with your own mixture of confits and cooked, raw, and marinated mushrooms. Think about shiitakes, chanterelles, button mushrooms, *Pleurotus*, etc.

Mushroom
curry quiche

Preparation: 30 min   Cooking time: 30–40 min

Serves 6

- ◉ 7 ounces (200 g) button mushrooms
- ◉ 7 ounces (200 g) slippery jack, chanterelle, or black trumpet mushrooms
- ◉ 2 teaspoons curry
- ◉ 1 puff pastry dough
- ◉ 3 eggs
- ◉ ½ cup (10 cl) milk
- ◉ ¾ cup (20 cl) heavy cream
- ◉ 1 tablespoon olive oil
- ◉ Salt and freshly ground pepper

Equipment
- ◉ Quiche or tart pan

1 Preheat the oven to 350°F (180°C, thermostat 6). In a pan, heat the olive oil and cook the mushrooms. Add the curry once all of the water has evaporated, then set aside.

2 Roll out the pastry dough and place it in the mold.

3 In a bowl, beat the eggs together with the milk and cream. Add the curried mushrooms. Season with salt and pepper then pour over the dough and bake for 30 to 40 minutes.

Stuffed
ceps

Preparation: 15 min   Cooking time: 35 min

Serves 4

- ◉ 4 large ceps
- ◉ 1 handful of chanterelles
- ◉ 11 ounces (300 g) sausage meat
- ◉ 1 egg
- ◉ 1 tablespoon breadcrumbs
- ◉ 1 garlic clove, chopped
- ◉ 2 tablespoons chopped parsley
- ◉ 2 slices of dry-cured ham
- ◉ 4 tomatoes, divided
- ◉ A little thyme
- ◉ 1 glass dry white wine
- ◉ Salt and freshly ground pepper

1 Remove the stems from the ceps and set aside. Roughly chop the ceps and sauté in a pan for 3 or 4 minutes with the chanterelles. Let cool.

2 In a bowl, combine the cep stems, sausage meat, egg, breadcrumbs, garlic, parsley, small slices of ham, one tomato cut into cubes, thyme, salt, and pepper.

3 Preheat the oven to 350°F (180°C, thermostat 6). Lightly oil a baking dish, slice the three remaining tomatoes, and place them at the bottom of the dish. Put the cep caps on top and stuff them with the filling mixture. Pour the white wine into the dish and add salt and pepper. Bake for 35 minutes.

Creamy chicken and
cep pappardelle

Preparation: 30 min   Cooking time: 20 min

Serves 4

Pasta
- 14 ounces (400 g) pappardelle

Sauce
- 10 ½ ounces (300 g) ceps
- 3 chicken breasts
- 2 garlic cloves
- ½ bunch flat-leaf parsley
- ¾ cup (20 cl) heavy cream
- 1 ½ tablespoons (20 g) butter
- 2 drizzles of olive oil, divided
- 1 drizzle white truffle oil
- 1 cup (100 g) freshly grated parmesan
- Salt and freshly ground pepper

1 Prepare the sauce. Scrub the ceps. Cut them into thick chunks. Cut the chicken breasts into strips. Peel the garlic cloves and remove any sprouts. Finely chop or press the garlic clove. Finely chop the parsley.

2 In a large pan, heat a drizzle of olive oil and cook the chicken strips over high heat. When they are golden brown, season with salt and pepper and set aside on a plate.

3 In the same pan, drizzle a little bit more olive oil and melt the butter. When it starts to foam, add the ceps and cook over high heat to reach a nice golden color. The water released from the mushrooms should be completely evaporated by the end of cooking. Season with salt and add the garlic and parsley. Add the chicken and the cream to the pan. Mix and continue cooking over a low flame for about 5 minutes.

4 Plunge the pasta into a pot of salted boiling water. When it is *al dente* (around 2 or 3 minutes), drain the pasta and pour it into the pan with the rest of the ingredients. Stir together for 1 minute over low heat. Drizzle with white truffle oil and serve with grated parmesan.

Prawns
en papillote

Preparation: 15 min   Cooking time: 20 min

Serves 4

- 18 prawns
- 12 button mushrooms
- 6 radishes
- ½ bunch cilantro
- 5 ¼ ounces (150 g) soy bean sprouts
- 1 lemon
- 2 ¾ tablespoons (4 cl) olive oil
- Salt and freshly ground pepper

Equipment

- 6 squares of parchment paper 10 x 10 inches (25 x 25 cm)

1 Preheat the oven to 410°F (210°C, thermostat 7).

2 Remove the prawn heads and peel the prawns completely. Remove the mushroom stems and finely dice the mushrooms along with the radishes. Strip half of the cilantro stems.

3 Place a portion of soy bean sprouts in the center of each parchment paper square. On top of that place three prawns on each square, then cover with the minced mushrooms and radishes.

4 Sprinkle with lemon juice and a drizzle of olive oil. Season with salt and pepper before sealing the parchment envelopes.

5 Bake for 15 minutes. After removing them from the oven, open your envelopes and sprinkle with cilantro leaves before serving.

Forester's chicken
en papillote

Preparation: 15 min   Cooking time: 45 min

Serves 6

- ⊙ 18 ounces (500 g) of a mixture of forest mushrooms (wood blewits, oyster mushrooms, *Russula*, slippery jacks, chanterelles, etc.)
- ⊙ 1 onion
- ⊙ ¾ cup (20 cl) heavy cream
- ⊙ 6 chicken legs
- ⊙ 2 ¾ tablespoons (4 cl) olive oil, divided
- ⊙ Salt and freshly ground pepper

Equipment

- ⊙ 6 squares of parchment paper 10 x 10 inches (25 x 25 cm).

1 Preheat the oven to 410°F (210°C, thermostat 7).

2 Remove the stems from the mushrooms and chop the mushrooms finely. Peel and chop the onion. In a pan, heat half of the olive oil. Cook the onion over low heat for 2 minutes.

3 Add the mushrooms and continue cooking over high heat until almost all of the liquid is evaporated. Add the cream, season with salt and pepper, and cook for 5 more minutes. In another pan, cook the chicken legs in the rest of the oil until they are golden brown.

4 Place one chicken leg in the center of each parchment paper square. Cover each one with the mushroom cream and seal the envelopes. Bake for 35 minutes.

Fish matelote stew
with white wine

Preparation: 30 min   Cooking time: 1 hour

Serves 6

Sauce
- 4 shallots
- 18 ounces (500 g) oyster mushrooms or button mushrooms
- 2 pats of butter
- 1 bottle (75 cl) white wine
- 2 cups (50 cl) fish stock (reconstituted, off the shelf)
- 2 cups (50 cl) thick crème fraîche
- 1 tablespoon flour
- 1 bunch chives
- Salt and freshly ground pepper

Vegetables
- 1 bunch green onions
- 1 bunch baby carrots
- 2 pats of butter
- 1 tablespoon sugar

Fish
- 4 ½ pounds (2 kg) fish (pike, zander, perch), gutted and scaled

1 Prepare the sauce. Peel and chop the shallots. Wash the mushrooms and cut them in half. In a pot, cook the shallots and mushrooms over low heat in the butter. Pour in the white wine. Bring to a boil and cook until the liquid has reduced by half.

2 Add the fish stock and bring to a boil again. During this time, carefully mix the crème fraîche with the flour. Add this mixture to the pot and stir well. Allow the pot to simmer over low heat for a few minutes.

3 Peel the green onions and carrots. Combine the vegetables, butter, sugar, and two glasses of water in a saucepan. Cover and cook for at least 15 minutes. Set aside.

4 Rinse and dry the fish well. Cut into slices 1.5–2 inches long (4–5 cm). Season with salt and pepper.

5 Add the fish and drained vegetables to the white wine sauce. Simmer for around 20 minutes without letting it boil. Add salt and pepper and then the chopped chives. Serve hot in bowls.

Grandmother's
chicken fricassée

Preparation: 25 min   Cooking time: 1 hour

Serves 4–6

- 1 free range chicken
- 10 ½ ounces (300 g) button mushrooms or pig's ear mushrooms
- 7 ounces (200 g) small green onions
- 2 shallots
- 1 bunch chives
- 5 ⅓ ounces (150 g) bacon cubes
- 2 glasses dry white wine
- 2 ½ cups (60 cl) chicken stock
- 1 pat of butter
- Sunflower oil
- Salt and freshly ground pepper

1 Break down the chicken into pieces and season. Clean the mushrooms and cut them in half. Peel the green onions. Peel and chop the shallots. Chop the chives.

2 In a pot, cook the chicken pieces in a little oil over high heat for 15 minutes until they are golden brown. Pour off the grease and add the bacon cubes, onions, mushrooms, and shallots.

3 Stir, cover, and cook for 5 min. Pour in the white wine and allow the liquid to boil 5 minutes before adding the chicken stock. Cover and cook for 40 minutes.

4 Use a ladle to remove the fat from the sauce, add the butter and the chopped chives. Stir and adjust seasoning if needed.

Seared beef
carpaccio

Preparation: 15 min   Freezing: 30 min   Cooking time: 3 min

Serves 4

- 18 ounces (500 g) filet of beef
- 1 bunch of basil
- 2 garlic cloves
- ¾ cup (100 g) pine nuts, divided
- 1 lemon
- 12 ¼ ounces (350 g) button mushrooms or chanterelles
- 8 cups (250 g) mesclun
- 3 ½ ounces (100 g) parmesan
- 6 tablespoons olive oil, divided
- Salt and freshly ground pepper

Equipment
- Blender (or food processor)

1 Tightly wrap the filet of beef in plastic wrap and put in the freezer for 30 minutes.

2 Remove the basil stems and rinse and dry the leaves. Peel and crush the garlic cloves. Blend all of this with three quarters of the pine nuts, half the olive oil, the juice from the lemon, salt, and pepper. Save for later.

3 Clean and dry the mushrooms, then mince them.

4 Remove the meat from the freezer and cut it into thin slices. Sear the slices for 2 minutes in a wok with the remaining oil over high heat, stirring. Add half of the sauce and the mushrooms and stir for 2 to 3 more minutes.

5 Serve the meat on a bed of mesclun seasoned with the rest of the sauce. Prepare thin parmesan shavings and place them on top of the meat. Sprinkle the whole dish with the remaining pine nuts and serve.

Option
You can also use thin veal or chicken cutlets; just cook them longer than the beef.

Cod and
mussel stew

Preparation: 25 min   Cooking time: 30 min

Serves 4

- 2 ¼ pounds (1 kg) mussels
- 2 shallots
- 5 ¼ ounces (150 g) button mushrooms
- ½ cup (10 cl) white wine
- 2 cups (50 cl) crème fraîche
- 1 flat tablespoon cornstarch
- 2 pinches of saffron threads
- 1 pat of butter
- 1 ¼ pounds (600 g) of cod fillets without skin or bones
- Salt and freshly ground pepper

Equipment
- Strainer

1 Cook the mussels in a covered pot for 8 to 10 minutes over medium heat. Take off of the heat and remove the shells. Filter through a strainer and save the cooking liquid. Peel and chop the shallots. Cut the mushrooms in thin slices.

2 In a pot, boil the mussel cooking liquid with the wine, shallots, and mushrooms for 4 or 5 minutes. As soon as half of the liquid has evaporated, stir together the crème fraîche and cornstarch and add this mixture to the pot. Bring to a boil, add the saffron and the butter. Remove from heat immediately. Season with salt and pepper.

3 Cut the cod into fairly large chunks or cubes. Season with salt and pepper. Add the cod and the mussels to the sauce, add salt and pepper, then cover and cook over low heat for around 20 minutes.

marengo

Preparation: 25 min   Cooking time: 1 hour 40 min

Serves 4–6

- 3 ¼ pounds (1.5 kg) veal shoulder
- 4 onions
- 2 garlic cloves
- ⅓ cup (40 g) flour
- 1 small bouquet garni
- ¾ cup (20 cl) white wine
- 1 tablespoon tomato paste
- 1 ¾ pounds (800 g) tomatoes
- 32 ounces (1 liter) veal stock
- 11 ounces (300 g) small button mushrooms
- 1 bunch chives
- Peanut oil
- Salt and freshly ground pepper

1 Remove connective tissue and cut the meat into pieces. Season with salt and pepper. Peel and chop the onions and garlic cloves.

2 In a large pot, heat 3 or 4 tablespoons of peanut oil over high heat. Then brown the meat pieces for around 10 minutes before sprinkling them with flour. Stir and continue cooking 2 or 3 minutes. Then add the chopped onion and garlic. Allow this to cook for 5 minutes, stirring from time to time.

3 Add the bouquet garni, white wine, tomato paste, sliced tomatoes, and veal stock. Cover and cook for about 1 hour.

4 Remove the fat from the sauce using a spoon then add the mushrooms after washing and cutting them in half. Continue cooking for another 30 minutes. Use a ladle to remove fat from the surface of the sauce again. Finally, add the chopped chives, salt, and pepper. Serve hot.

Morel

sauce

Preparation: 10 min   Cooking time: 15 min

Serves 6

- 2 shallots
- 3 ½ ounces (100 g) fresh morels
- A few chive stems
- 1 ¾ tablespoons (25 g) butter
- 6 ¾ tablespoons (10 cl) white wine
- 6 ¾ tablespoons (10 cl) chicken stock
- 1 cup (25 cl) heavy cream
- Salt and freshly ground pepper

1 Peel and chop the shallots. Cut the morels in half vertically and clean them under cold water to get rid of sand. Chope the chives as finely as possible.

2 In a pan, melt the butter and cook the chopped shallots with the morels over medium heat. When the shallots are translucent, pour in the wine and chicken stock, then reduce until almost all of the liquid is gone.

3 Next add the cream and bring to a boil. Reduce again until the liquid has reduced by half in order to obtain a thick sauce. Salt and pepper to taste. Sprinkle with chives and serve immediately.

Note

This sauce is delicious with grilled white meats (veal or chicken) or over green asparagus accompanied by poached eggs and salmon roe.

Suggestion

You can also make this recipe using dehydrated morels. Soak them in water for a few minutes before using.

Pork boulettes
with poppy seeds

Preparation: 15 min   Cooking time: 20 min

Serves 4

- ½ bunch flat-leaf parsley
- 1 large onion
- 26 ½ ounces (750 g) lean pork shoulder
- 2 tablespoons flour
- 2 tablespoons poppy seeds
- Salt and freshly ground pepper

Sauce

- 12 ¼ ounces (350 g) slippery jacks or bronze boletes
- 4 tablespoons peanut oil, divided
- 1 cup (25 cl) vegetable broth
- 1 teaspoon paprika
- 1 cup (25 cl) whipping cream
- Salt and freshly ground pepper

1 Remove the parsley stems, rinse, and chop finely. Peel and chop the onion. Chop the meat. Mix these ingredients together. Add salt and pepper. Form balls the size of walnuts.

2 Roll the boulettes in the flour, then in the poppy seeds, pressing to make them stick. Heat half of the oil in a wok and cook the boulettes until browned. Remove from wok and set aside.

3 Prepare the sauce. Clean and mince the mushrooms. Add the rest of the oil to the wok. Pour in the mushrooms along with salt and pepper, then sauté until they have released all of their liquid. Return the boulettes to the wok, add the broth, and cook for 10 minutes. Add the paprika and whipping cream and adjust seasoning if needed.

Lamb stew
with tomatoes

Preparation: 25 min   Cooking time: 2 hours

Serves 4–6

- 4 ½ pounds (2 kg) lamb shoulder
- 4 onions
- 4 garlic cloves
- 1 ¾ pounds (800 g) tomatoes
- 11 ounces (300 g) button mushrooms, Scotch bonnets, or St. George's mushrooms
- 1 bunch chives
- ⅓ cup (40 g) flour
- 1 small bouquet garni
- ¾ cup (20 cl) white or rosé wine
- 1 tablespoon tomato paste
- 1 sprig fresh thyme
- 2 cups (50 cl) chicken stock
- 5 tablespoons peanut oil
- 4 tablespoons olive oil
- Salt and freshly ground pepper

1 Break down the lamb shoulder into pieces and season. Peel and chop the onions and the garlic cloves. Slice the tomatoes. Clean the mushrooms. Chop the chives.

2 In a large pot, heat the peanut oil over a high flame. Brown the meat pieces for around 10 minutes, then sprinkle with flour. Stir and continue cooking for 1 or 2 minutes. Add the onions and the garlic. Cook for 5 more minutes, stirring occasionally.

3 Add the bouquet garni, the wine, tomato paste, tomato slices, thyme, and chicken stock. Cover and simmer for about 1 hour.

4 Remove the fat from the top of the sauce using a spoon, then add the mushrooms and continue cooking for 40 more minutes. Finish by adding the chives and olive oil. Season with salt and pepper. Serve hot.

Suggestion
When cooking has finished, add black and green olives, basil, a few capers, and a little fresh chili pepper to your lamb stew. It can also be garnished with chopped sundried tomatoes.

Mini mushroom
gratins

Preparation: 25 min   Cooking time: 40 min

Serves 2

- 18 ounces (500 g) of various mushrooms (saffron milkcaps, black trumpets, gypsy mushrooms, king boletes, etc.)
- 1 chopped shallot
- 2 eggs
- 1 ¼ cup (30 cl) heavy cream
- 2 tablespoons grated parmesan
- 1 pat of butter
- 1 tablespoon olive oil
- Freshly grated nutmeg
- Salt and freshly ground pepper

Equipment
- 2 mini gratin dishes

1 Clean and dry off the mushrooms. If they are too large, cut them in half.

2 In a pan, cook the mushrooms and shallots in the hot butter and oil. Cook for 10 to 15 minutes. All of the water from the mushrooms must be evaporated.

3 Season the mushrooms and divide them between the mini gratin dishes. In a bowl, beat together the eggs, cream, salt, pepper, and nutmeg. Pour over the mushrooms, sprinkle with parmesan, and bake in the oven for 20 minutes at 375°F (190°C, thermostat 6).

Slow-cooked
escargot with ceps

Preparation: 25 min   Cooking time: 25 min

Serves 6 (appetizers)

- 4 ¼ ounces (120 g) smoked bacon cubes
- 2 garlic cloves
- A few sprigs of parsley
- 18 ounces (500 g) ceps
- 48 canned snails, drained
- 4 tablespoons thick crème fraîche
- ¾ cup (20 cl) broth
- 1 pat of butter
- Oil
- Salt and freshly ground pepper

Equipment
- Skimmer

1 Bring a pot of water to boil. Blanch the bacon cubes, then drain them and set them aside. Peel the garlic cloves, remove the sprout if needed, then chop them. Finely chop the parsley (except for the larger sprigs).

2 In a pan, sauté the ceps for 10 minutes with a little oil, salt, and pepper. Set aside. Preheat the oven to 350°F (180°C, thermostat 6).

3 Melt the butter in a small pot over high heat. Sauté the bacon, ceps, and snails for 2 or 3 minutes, roughly. Add the garlic and cook for another minute.

4 Pour in the crème fraîche and the broth. Season with salt and pepper and sprinkle in the parsley. Cover and cook in the oven for 15 to 20 minutes. Serve hot.

Option
Serve these escargots with the ceps in puff pastry shells.

Goat cheese-stuffed
mushrooms

Preparation: 10 min   Cooking time: 10–15 min

For 6 mushrooms

- 6 large mushrooms (saffron milkcaps, ceps, or large bolets)
- ½ log goat cheese
- 1 tablespoon crème fraîche
- 1 tablespoon breadcrumbs
- 10 chive stems
- 3 sprigs parsley

Equipment
- Mixer

1 Preheat the oven to 350°F (180°C, thermostat 6). Remove the mushroom stems and wash the caps well.

2 In the bowl of a mixer, blend together the goat cheese, crème fraîche, breadcrumbs, chives, and parsley. Once all of the ingredients are mixed in, spoon the filling into the mushroom caps. Place the filled caps on a baking sheet.

3 Bake the stuffed mushrooms for 10 to 15 minutes and serve hot.

Suggestion
You can prepare the mushrooms a few hours ahead and cook them at the last minute.

Black truffle
tagliolini

Preparation: 30 min   Cooking time: 8 min

Serves 4

Pasta
- 14 ounces (400 g) tagliolini

Sauce
- 2 ounces (60 g) black truffles
- 1 garlic clove
- ½ cup (50 g) freshly grated grana padano
- 6 tablespoons olive oil
- Salt

Equipment
- Vegetable peeler

1 Scrub the truffle under running water. Use a vegetable peeler to shave it into fine slices, keeping the small pieces and chopping them. Peel the garlic and crush it with the flat side of a knife. In a pan, heat the olive oil with the garlic and small truffle pieces. When the garlic begins to brown, remove it from the pan and turn off the flame.

2 Plunge the tagliolini into a pot of salted boiling water. As soon as they are cooked *al dente*, drain the pasta and pour it into the other pan. Mix with the oil and truffle pieces then add the truffle shavings. Serve with the grated grana padano.

Mixed
mushroom ragoût

Preparation: 15 min   Cooking time: 35 min

Serves 4

- 5 shallots
- 7 ounces (200 g) ceps
- 7 ounces (200 g) chanterelles
- 7 ounces (200 g) black trumpets
- 7 ounces (200 g) shiitakes
- 7 ounces (200 g) button mushrooms
- 1 pat of butter
- 3 ½ tablespoons (5 cl) port wine
- 1 ⅔ cups (40 cl) light crème fraîche
- 1 bunch chervil
- 1 pinch powdered nutmeg
- 1 tablespoon parmesan (optional)
- Salt and freshly ground pepper

1 Peel and finely chop the shallots. Clean and cut the largest mushrooms if needed.

2 In a pot, heat the butter over medium heat then add the shallots and mushrooms. Cook for 5 minutes, stirring occasionally. Add the port wine and cook for another 5 minutes.

3 Add the crème fraîche, season with salt and pepper, then let simmer for 25 minutes. When cooking is finished, add additional salt and pepper if needed. Add the chervil, nutmeg, and parmesan if desired. Serve hot.

Suggestion

Replace half of the crème fraîche with an equivalent amount of chicken stock.
You can also serve this ragoût in puff pastry shells. If the sauce is too runny, thicken it with a little cornstarch.

Guinea fowl
with morels and vin jaune

Preparation: 20 min   Cooking time: 1 hour

Serves 6

- 1 onion
- 1 guinea fowl, cut into pieces
- 5 ¼ ounces (150 g) morels
- 1 ¼ cups (30 cl) vin jaune
- 2 ½ ounces (70 g) butter, divided
- 1 ¼ cups (30 cl) heavy cream
- Salt and freshly ground pepper

1 Peel and chop the onion.

2 In a pot, cook the guinea fowl pieces in roughly half of the butter and seasoned with salt and pepper. Add the onion and cook for a few minutes. Add the vin jaune. Cover the pot, reduce the heat, and continue cooking for 40 minutes.

3 In a pan, cook the morels in the rest of the butter for 10 minutes over medium heat. Add them to the guinea fowl and pour in the cream. Continue cooking for 10 minutes and serve.

Cream of mushroom soup
with mustard

Preparation: 15 min   Cooking time: 20 min

Serves 6

- 1 ½ pounds (700 g) button mushrooms
- 3 tablespoons (10 g) ceps or black trumpets
- 3 garlic cloves
- 2 ounces (60 g) butter, divided
- ⅔ cup (15 cl) heavy cream
- 2 tablespoons strong mustard
- Salt and freshly ground pepper

Equipment

- Hand mixer

1 Cut the mushrooms into quarters and wash them under cold water. In a pot, sauté the chopped garlic in half of the butter for 2 minutes. Add the mushrooms, salt, and pepper, then cover with cold water. Cover the pot three quarters of the way and leave to cook for 20 minutes over medium heat.

2 Using a hand mixer, blend the contents of the pot to obtain a smooth and creamy texture. Add the cream and the rest of the butter and mix a little more.

3 Return the creamy soup to heat over a low flame, and when you are ready to serve it, incorporate the mustard. Stir without allowing the soup to boil and serve immediately.

Suggestion
Accompany this creamy soup with small toasted croutons rubbed with garlic and sprinkled with olive oil.

Mushroom
parmesan tian

Preparation: 15 min   Cooking time: 15 min

Serves 4

- ◉ 18 ounces (500 g) yellow-leg chante-relles or button mushrooms
- ◉ 3 ½ tablespoons (30 g) shelled hazelnuts
- ◉ ⅔ cup (15 cl) whipping cream
- ◉ 1 cup (100 g) grated parmesan
- ◉ Salt and freshly ground pepper

1 Preheat the oven to 425°F (220°C, thermostat 7-8). Cut the mushrooms into very thin slices using a small, sharp knife. Roughly crush the hazelnuts.

2 Arrange the mushrooms in a gratin dish. Cover with cream and parmesan. Add salt and pepper and sprinkle the crushed hazelnuts on top. Cook for 15 minutes at 425°F (220°C, thermostat 7–8) until the tian is golden brown.

Sautéed mushrooms
in persillade sauce

Preparation: 10 min   Cooking time: 6–10 min

Serves 4–6

- ◉ 2 garlic cloves
- ◉ 10 sprigs flat-leaf parsley
- ◉ 1 pat of butter
- ◉ 7 ounces (200 g) ceps
- ◉ 7 ounces (200 g) chanterelles
- ◉ 7 ounces (200 g) black trumpets
- ◉ 7 ounces (200 g) *Craterellus tubaeformis*
- ◉ 7 ounces (200 g) button mushrooms
- ◉ A few drops of lemon juice
- ◉ Salt and freshly ground pepper

1 Peel and chop the garlic cloves. Also chop the parsley. In a large pan, heat the butter over medium heat then add all of the mushrooms once they have been cleaned.

2 Season with salt and pepper and sauté 5 or 6 minutes, stirring often.

3 Add the parsley and garlic mixture, stir together, and cook for 2 or 3 more minutes. Salt and pepper to taste, then add a few drops of lemon juice and serve.

Note

As they cook, mushrooms release a strongly scented water. This liquid should not be discarded and should instead be allowed to evaporate in the pan. Depending on the mushrooms, the size of the pan, and the intensity of the flame, cooking time may vary. You can serve the mushrooms with a little bit extra of this pleasant-smelling liquid.

Sautéed cod with aromatic rice and
black mushrooms

Preparation: 20 min   Cooking time: 30 min

Serves 4

- ◉ 4 tablespoons dried black mushrooms
- ◉ 4 green onions
- ◉ 2 garlic cloves
- ◉ 1 red chili
- ◉ 1 ¼ pounds (600 g) cod
- ◉ ¾ cup (150 g) aromatic rice
- ◉ 5 tablespoons teriyaki sauce
- ◉ 3 tablespoons flour
- ◉ 7 ounces (200 g) bamboo shoots
- ◉ 5 tablespoons sesame oil, divided
- ◉ Salt and freshly ground pepper

1 Soak the mushrooms in a bowl of hot water for 15 minutes. Drain. Clean the onions, peel the garlic cloves, and remove the chili pepper seeds. Chop all of these ingredients finely. Dry the fish and cover it in a mixture of flour, salt, and pepper.

2 Rinse, dry, and cut the fish into large pieces. Sprinkle with half of the teriyaki sauce and keep at room temperature.

3 Steam the rice in twice its volume of water for 15 minutes. Let cool and fluff with a fork.

4 Fry the fish in a wok in half of the sesame oil. Remove it and clean the wok with a paper towel. In the clean wok, sauté the onions, chili, mushrooms, drained bamboo shoots, rice, and garlic cloves in the remaining oil. Add the rest of the teriyaki sauce and the fish. Add salt and pepper before serving.

Option

Replace the cod with pollock or whiting and vary the vegetables depending on your tastes.

Eggplant and
Spicy Mushroom Tart

Preparation: 35 min   Cooking time: 30 min

Serves 6

- 1 puff pastry dough
- 2 eggplants
- ½ onion
- 3 ½ ounces (100 g) cleaned button mushrooms or orange boletes
- ½ teaspoon curry powder
- ½ teaspoon sweet pepper
- 1 tablespoon pine nuts
- Olive oil

Equipment
- Tart pan

1 Preheat the oven to 350°F (180°C, thermostat 6). Roll out your puff pastry dough and place it in the tart pan.

2 Peel the eggplants and onion and cut into cubes. Cook them in a well-oiled pan with the mushrooms. Sprinkle the spices over the pan and cook for 20 minutes.

3 Spread the mixture over your puff pastry, sprinkle on the pine nuts, and bake for 30 minutes.

Ceps
en papillote

Preparation: 10 min   Cooking time: 20 min

Serves 6

- ⊙ 2 ¼ pounds (1 kg) ceps
- ⊙ 8 tablespoons (12 cl) olive oil
- ⊙ 6 bay leaves
- ⊙ Salt and freshly ground pepper

Equipment
- ⊙ 6 squares of parchment paper 10 x 10 inches (25 x 25 cm)

1 Preheat the oven to 350°F (180°C, thermostat 6).

2 Scrub the ceps, cut off the earthy stems, and rinse quickly under the tap. Slice them in large chunks and place in a bowl. Drizzle with olive oil and season with salt and pepper.

3 Divide the ceps between the 6 pieces of parchment paper. Add one bay leaf to each envelope and close them. Cook for 20 minutes and serve the envelopes as soon as they come out of the oven.

mushrooms

Preparation: 25 min   Cooking time: 15 min   Refrigeration: 48 hours

Serves 4

- ⊙ 4 ceps
- ⊙ 8 large button mushrooms
- ⊙ 4 red onions
- ⊙ A few sprigs of thyme
- ⊙ 20 basil leaves
- ⊙ ¾ cup (20 cl) rice vinegar, white wine vinegar, or apple cider vinegar
- ⊙ 1 ⅔ cups (40 cl) grapeseed oil
- ⊙ Peanut oil
- ⊙ Salt and freshly ground pepper

1 Cut the mushrooms in thick slices. Peel the onions and slice them into thick rounds. Season with salt and pepper. Place the mushrooms on a hot oiled grill. Allow to cook for 3 or 4 minutes then turn them over and cook for another 2 or 3 minutes. Cook the onions the same way.

2 Place all of the mushrooms and onions in a bowl. Season once more with salt and pepper then sprinkle with the thyme and basil. Finally, pour over the vinegar and grapeseed oil. Mix well. Marinate for 48 hours in the refrigerator, then enjoy.

Suggestion
Drain the mushrooms and eat them alone, with an arugula salad, in pasta, or on toast.

---

duck breast tartare

Preparation: 30 min   Refrigeration: 30 min

Serves 4

- ⊙ 8 ½ ounces (250 g) smoked duck breast
- ⊙ 14 ounces (400 g) button mushrooms
- ⊙ 2 shallots
- ⊙ 1 bunch chives
- ⊙ 2 tablespoons capers
- ⊙ 6 cornichon pickles
- ⊙ 3 tablespoons sunflower oil
- ⊙ 2 tablespoons sherry vinegar
- ⊙ Salt and freshly ground pepper

Accompaniment
- ⊙ 8 cups (250 g) mesclun and a vinaigrette made with 3 tablespoons olive oil
- ⊙ 1 dash of mustard
- ⊙ 1 tablespoon vinegar
- ⊙ Salt and freshly ground pepper

Equipment
- ⊙ 4 small aluminum trays or 4 ramekins

1 Chop the pickles and the capers. Peel the shallots and cut them in small cubes. Clean and chop the chives. Clean the mushrooms and remove the fat around the duck breast.

2 Thinly slice the duck breast and the mushrooms. Add the pickles, capers, chives, and shallots. Pour in the oil and the sherry vinegar. Season with salt and pepper and mix well.

3 Fill the small aluminum trays or ramekins with the tartare, packing them to the brim, and refrigerate for 30 minutes.

4 When it is time to serve, toss the mesclun with the vinaigrette and divide between the plates. Remove each tartare from its mold and place in the center of each plate. Serve with toasted bread.

Tagliatelle with
chanterelles and chicken

Preparation: 20 min   Cooking time: 1 hour

Serves 4

- ◉ 1 ¼ pounds (600 g) fresh tagliatelle
- ◉ 1 ¾ pounds (800 g) chanterelles
- ◉ 1 carrot
- ◉ 1 shallot
- ◉ 1 free range chicken weighing 3 ¼ pounds (1.5 kg)
- ◉ 6 ¾ tablespoons (10 cl) dry white wine
- ◉ 1 organic chicken bouillon cube
- ◉ ⅔ cup (15 cl) heavy cream
- ◉ 4 tablespoons olive oil
- ◉ Salt and freshly ground pepper

1 Clean the chanterelles. Rinse them quickly under running water. Peel and chop the carrot and shallot in small cubes. Heat the oil in a pot and brown the chicken on each side after seasoning it.

2 Remove the chicken from the pot and add the carrot and the shallot. Cook for 2 minutes, then add the white wine. Reduce the liquid by a third and add roughly seven tablespoons (10 cl) water, the bouillon cube, the heavy cream, chicken, and mushrooms. Cover the pot and cook for 30 minutes over medium heat.

3 Remove the cover from the pot and allow the liquid to reduce for 15 minutes. The chicken is cooked when you pierce the thighs with a fork and the liquid that runs out is transparent. Check the seasoning and keep warm while you cook the pasta. Serve quickly.

Suggestion

You can prepare this dish with chicken thighs instead, but the whole chicken gives more flavor to the sauce and lets each person find their favorite little piece.

Tagliatelle with
chanterelles and chicken

Preparation: 20 min   Cooking time: 1 hour

Serves 4

- ⊙ 1 ¼ pounds (600 g) fresh tagliatelle
- ⊙ 1 ¾ pounds (800 g) chanterelles
- ⊙ 1 carrot
- ⊙ 1 shallot
- ⊙ 1 free range chicken weighing 3 ¼ pounds (1.5 kg)
- ⊙ 6 ¾ tablespoons (10 cl) dry white wine
- ⊙ 1 organic chicken bouillon cube
- ⊙ ⅔ cup (15 cl) heavy cream
- ⊙ 4 tablespoons olive oil
- ⊙ Salt and freshly ground pepper

1 Clean the chanterelles. Rinse them quickly under running water. Peel and chop the carrot and shallot in small cubes. Heat the oil in a pot and brown the chicken on each side after seasoning it.

2 Remove the chicken from the pot and add the carrot and the shallot. Cook for 2 minutes, then add the white wine. Reduce the liquid by a third and add roughly seven tablespoons (10 cl) water, the bouillon cube, the heavy cream, chicken, and mushrooms. Cover the pot and cook for 30 minutes over medium heat.

3 Remove the cover from the pot and allow the liquid to reduce for 15 minutes. The chicken is cooked when you pierce the thighs with a fork and the liquid that runs out is transparent. Check the seasoning and keep warm while you cook the pasta. Serve quickly.

Suggestion
You can prepare this dish with chicken thighs instead, but the whole chicken gives more flavor to the sauce and lets each person find their favorite little piece.

Rump steak and
mushroom pudding

Preparation: 1 hour   Resting time: 1 hour 15 min   Cooking time: 2 hours

Serves 8–10

- 1 pâte brisée dough

Filling

- 1 ¼ pounds (600 g) rump steak
- 18 ounces (500 g) cleaned monk's head mushrooms
- 2 white onions
- 4 tablespoons (30 g) flour
- 1 ⅔ cups (40 cl) beef or chicken stock
- 1 egg
- Sunflower oil, butter
- Salt and freshly ground pepper

Equipment

- Mixer (optional)
- Pudding tin (4 ½–6 ⅔ pounds, or 2–3 kg)
- Aluminum foil

1 Cut the meat into cubes about one third of an inch (1 cm) in size. Cut the mushrooms in fairly thick slices. In a pot, brown the seasoned meat with the mushrooms and onions for about 5 minutes in the sunflower oil.

2 Add the flour, stir, and cook for 2 or 3 minutes before adding simmering broth. Cook for a few minutes. The mixture should be fairly thick. Season with salt and pepper, then remove from heat and let cool.

3 Preheat the oven to 300°F (150°C, thermostat 5). Roll out the dough and lay it in a generously buttered pudding tin, taking care to allow enough dough to spill over the edge to cover the filling. Spoon the filling into the pudding tin and cover with the pastry. Seal the edges of the pastry firmly. Using a small knife, make a hole in the top of the pudding and insert a small aluminum foil tube. Cover with a layer of greased aluminum foil. Bake for 2 hours. Allow the pudding to cool for 15 minutes outside the oven before serving.

Index

The boldface type indicates species description.

Edible Mushroom Habitats

Recipe Credits

The name of the author is followed by the photographer's where appropriate.

Photography Credits